Environmental Applications of Magnetic Nanomaterials Under the Influence of Magnetic Fields

Online at: https://doi.org/10.1088/978-0-7503-6377-8

Environmental Applications of Magnetic Nanomaterials Under the Influence of Magnetic Fields

Uyiosa Osagie Aigbe

Center for Space Research, North-West University, Private Bag X6001, Potchefstroom, 2531, South Africa

and

National Institute of Theoretical and Computational Science, Johannesburg, 2000, South Africa

and

School of Chemistry & Physics, University of KwaZulu-Natal, Pietermaritzburg Campus, Private Bag X01, Scottsville 3209, South Africa

Kingsley Eghonghon Ukhurebor

Department of Physics, Faculty of Science, Edo State University, Iyamho, P.M.B. 04, Auchi 312101, Edo State, Nigeria

IOP Publishing, Bristol, UK

ISBN   978-0-7503-6377-8 (ebook)
ISBN   978-0-7503-6373-0 (print)
ISBN   978-0-7503-6374-7 (myPrint)
ISBN   978-0-7503-6376-1 (mobi)

DOI   10.1088/978-0-7503-6377-8

Version: 20250401

IOP ebooks

British Library Cataloguing-in-Publication Data: A catalogue record for this book is available from the British Library.

Published by IOP Publishing, wholly owned by The Institute of Physics, London

IOP Publishing, No.2 The Distillery, Glassfields, Avon Street, Bristol, BS2 0GR, UK

US Office: IOP Publishing, Inc., 190 North Independence Mall West, Suite 601, Philadelphia, PA 19106, USA

To our families: past, present, and future.

Contents

Preface

Natural chemicals, heavy metals, dyes, pharmaceuticals, radioactive contaminants, and pathogens are among the contaminants being released into water bodies as a result of human and natural activities. In addition to threatening the availability of clean water in the future, this could destroy living organisms for a long time, alter their DNA, and even cause them to die. Hence, it is urgently necessary to advance water and/or wastewater treatment technology that is efficient, dependable, and affordable. Magnetic nanoparticles have had the biggest effect to date in the treatment of toxins in groundwater and marine environments. The adoption of magnetic nanoparticles in various fields has increasingly led to the inclusion of cutting-edge nanotechnologies in environmental, medical/biomedical, and therapeutic applications. As a result, this book titled *Environmental Applications of Magnetic Nanomaterials Under the Influence of Magnetic Fields* provides a foundation for the fabrication, characterization, development, utilization, and applications of magnetic nanomaterials in magnetic resonance imaging as well as for resolving environmental issues in society globally.

The book is made up of seven chapters, which primarily focus on the advancement of nanotechnology in the synthesis of novel magnetic nanomaterials and their extensive application in magnetic resonance imaging, as well as in various environmental applications. Also discussed in this book are the challenges associated with the use of magnetic nanomaterials in various fields of application. Hence, this book will serve as a reference and will be of great assistance to students, professionals, practitioners, scientists, researchers, and academicians in various research domains, particularly those in the environmental and materials science fields.

Acknowledgments

First and foremost, we are most grateful to Almighty God, our creator, who kept us alive and gave us all we needed to make this book a success. We owe Him everything. We have always been, and will forever be, grateful to Him.

Our profound gratitude goes to the managements of our respective institutions, who have always provided great assistance in our research and academic development. We are also grateful to the authors and publishers whose publications were used as a basis for writing this book, without whom this book would not have been possible.

We thank our families, mentors, colleagues, friends, and all who contributed and are still contributing, in one way or another, toward the success of our academic and research endeavors as well as our other pursuits. May God bless you all richly!

Author biographies

Uyiosa Osagie Aigbe

Uyiosa Osagie Aigbe is a former research fellow with the Department of Mathematics and Physics, Faculty of Applied Science, Cape Peninsula University of Technology, Cape Town, South Africa, and currently a research associate with the Center for Space Research, North-West University, Potchefstroom, South Africa, the National Institute of Theoretical and Computational Science, Johannesburg, South Africa as well as the School of Chemistry & Physics, University of KwaZulu-Natal, Pietermaritzburg Campus, Scottsville, South Africa. He obtained his PhD degree in physics from the prestigious University of South Africa, Pretoria, South Africa. He is currently a member of several learned academic organizations. His research interests are in applied physics, nanotechnology, fluid dynamics, water purification processes, image processing, environmental physics, machine learning, statistical analysis, and materials science. He has also served as a reviewer and editor for numerous highly regarded publishers, such as Elsevier, Springer Nature, the Royal Society of Chemistry (RSC), the Institute of Physics (IOP), Taylor & Francis, Frontiers Media SA, MDPI, and Hindawi, and has authored or coauthored several research publications. His hobbies are reading, swimming, and playing/watching football. He is a Christian and is happily married with a family.

Kingsley Eghonghon Ukhurebor

Kingsley Eghonghon Ukhurebor is a senior lecturer/researcher and the immediate past acting head of the Department of Physics and the current acting director of sports at Edo State University, Iyamho, Nigeria. He is also a research fellow at the West African Science Service Centre on Climate Change and Adapted Land Use (WASCAL), Competence Centre, Ouagadougou, Burkina Faso, a climate institute sponsored by the Federal Ministry of Education and Research, Germany. He was awarded his PhD in physics electronics by the University of Benin, Benin City, Nigeria; his MSc in physics electronics by the University of Benin, Benin City, Nigeria; his PGD in education by Usmanu Danfodiyo University, Sokoto, Nigeria; and his BSc in applied physics by Ambrose Alli University, Ekpoma, Nigeria. He is a member of several learned academic organizations, such as the Nigerian Young Academy (NYA), the Nigerian Institute of Physics (NIP), the Materials Science and Technology Society of Nigeria (MSN), the Teacher Registration Council of Nigeria (TRCN), etc. His research interests include applied physics, climate physics, environmental physics, telecommunications physics, machine learning, statistical analysis, and materials science (nanotechnology). He serves as an editor and reviewer for several reputable journals and publishers, including Springer Nature, Elsevier, the Royal Society of Chemistry (RSC), the Institute of Physics (IOP), Taylor & Francis, John Wiley & Sons, the IEEE, Frontiers Media SA, Hindawi, etc. and he acts as a supervisor and examiner for undergraduate and postgraduate students. He has authored or coauthored several publications with these reputable journals and publishers. He is currently ranked among the top 50 authors in Nigeria by Scopus scholarly output and is listed among the top 2% of scientists in the world by Stanford University, USA, and Elsevier. His hobbies include reading and playing/watching football. He is a Christian and is happily married with a family.

IOP Publishing

Environmental Applications of Magnetic Nanomaterials
Under the Influence of Magnetic Fields

Uyiosa Osagie Aigbe and Kingsley Eghonghon Ukhurebor

Chapter 1

Introduction to magnetic nanomaterials and the development of magnetic nanomaterials for environmental remediation

Researchers, scientists, and technologists have recently shown a great deal of interest in developing nanomaterials (NMs) that are inexpensive to produce and safe for the environment. There is vast potential for the exploration of ferrite NMs for utilization in environmental restoration, remediation, and management owing to these materials' distinctive magnetic characteristics, which can be modified to yield an enormous amount of functionalization. Hence, this chapter, which serves as an introduction to the book titled *Environmental Applications of Magnetic Nanomaterials Using Magnetic Fields*, will mainly focus on highlighting the contemporary applications and development of magnetic NMs (MNMs) for environmental remediation.

1.1 Introduction

NMs, which can be manufactured from a variety of materials and range in size from 1 to 100 nm, are the core focus of nanotechnology (NT) or nanoscience (NS) vis-à-vis biotechnology (BT)/nanobiotechnology (NBT). NT is a relatively new approach that is attracting interest for use in industrial, biological, scientific, technological, electronic, and environmental applications, as well as almost, if not all, aspects of human endeavor [1–5]. Due to NMs' highly specific and distinct qualities compared to their bulk forms, they are utilized, employed, and/or applied in a wide range of fields, including the field of biotechnological environmental remediation and the fabrication of devices and sensors. Their geometric properties, including shape, composition, homogeneity, and aggregation, are taken into account in their classification. Furthermore, nanostructures can be zero-dimensional (0D), one-dimensional (1D), two-dimensional (2D), or three-

doi:10.1088/978-0-7503-6377-8ch1

dimensional (3D) (such as nanotubes and nanowires) and can have a spherical, flat, needle-like, or random arrangement [2, 3, 6–8]. NMs exhibit unique properties such as optical, electrical, catalytic, and thermal characteristics. MNMs are one of the most essential, important, and significant categories of NMs, offering exceptional super-paramagnetism, a large surface area, and biocompatibility [5].

NMs can be produced in many different morphologies and are mostly found in powder, colloidal particles, and suspension phases [9]. Composites are materials that are combined with one or more other materials to form NMs. The constituents of composites include metals, alloys, ceramics, and polymers. Single-material composites include materials such as alloys, metals, polymers, and ceramics [10]. Numerous techniques, including coprecipitation, vapor deposition synthesis, gas-phase procedures, mechanical operations, and effective dispersal in a matrix, can be used to obtain their chemical and electrostatic characteristics [11]. Chemical synthesis procedures are frequently preferred over other approaches in order to obtain sufficient consistency. A variety of categories, including mono-metallic, bimetallic, trimetallic, metal oxide, magnetic hybrid, semiconductor, composite, and others, can be used to categorize NMs according to their composition [12]. However, the main topic of this chapter is MNMs, which are becoming more and more prevalent in a wide range of industries, such as BT/NBT and environmental remediation [11, 13–21]. MNMs have numerous applications; each MNM has a unique set of applications and benefits and thus particular advantages. Supercapacitors, as well as batteries (such as fuel cells and solar cells), biosensors (BSs)/nanobiosensors (NBSs), antibacterial materials, and smart memory products are only a few of the many products that use them [22]. Furthermore, advancements in point-of-care analysis/diagnosis, environmental cleanup, agricultural activities, and food safety have recently made use of bioactive proteins and MNMs [23]. In the field of technological engineering, it has long been known that the use of 'magneto-ferric oxide' and other magnetic elements in concrete increases both sustainability and durability [24]. The fundamental use and development of MNMs in NT/BNT for environmental remediation is also highlighted in this chapter [25].

As already stated at the start, this chapter, which serves as an introduction to this book titled *Environmental Applications of Magnetic Nanomaterials Using Magnetic Fields*, will mainly focus on highlighting the contemporary applications and development of MNMs for environmental remediation. Hence, the book aims to provide a foundation for the development, utilization, and applications of MNMs for resolving environmental issues in society globally. The other chapters of this book are structured as follows: the fabrication and characterization of MNMs and their application in the removal of contaminants under the effect of a magnetic field (MF) are discussed in chapter 2. Chapter 3 deals with MF imaging using MNMs. Chapter 4 highlights the applications of MFs for environmental sustainability using MNMs. Chapter 5 covers the application of MNMs in soil decontamination and the bioremediation of contaminants under the influence of an MF; chapter 6 discusses the effect of MFs in gas and contaminant sensing using MNMs; and chapter 7, which is the concluding chapter, discusses the challenges and prospects of MNMs. Figure 1.1 graphically outlines the various chapters that make up this book.

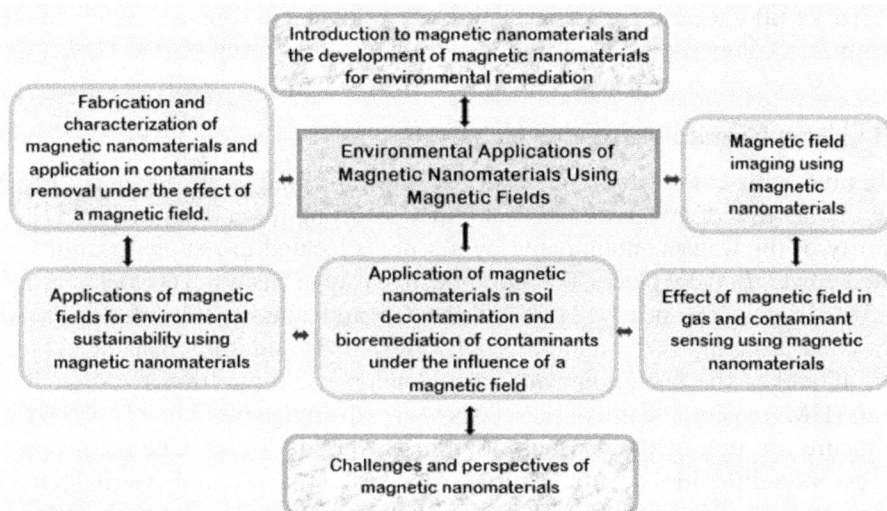

Figure 1.1. Graphical representation of the various chapters that make up this book.

1.2 A brief historical perspective on NMs

The first humans used fire, which resulted in the emergence of nanoscale particles of smoke [26]. However, NMs were not discovered until much later in the annals of science. One of the first scientific papers on this topic was Michael Faraday's 1857 publication of his research on the production of colloidal gold particles [9, 26]. More recently, researchers have been studying nanostructured catalysts for almost 80 years. In the early 1940s, precipitated silica NMs and fumed silica NMs were produced and commercialized (in industrialized form) in the US and Germany, respectively, as substitutes for ultrafine carbon black in rubber reinforcement [27]. Researchers produced metallic nanopowders in the 1960s and 1970s for magnetic recording cassettes, which were subsequently made available for purchase. Granqvist and Buhrman [28] published their first study of nanocrystals in 1976, which was hailed as innovative and employed the now commonly used inert gas evaporation process [28]. Some recent studies have revealed that Maya blue paint is a hybrid material or composite with a nanostructure that has numerous possibilities. The composite is made up of needle-shaped palygorskite (clay) crystals that combine or integrate to form superlattices with phases as small as 1.40 nm, despite the fact that genuine samples from Jaina Island (in the present-day Mexican state of Campeche) are still not known as contained in a report by [29].

With the application of nanophase engineering, we are now able to modify the mechanical, optical (optical–electronic), catalytic (electrical and magnetic), and electrical characteristics of an increasing range of structural as well as functional materials, primarily inorganic and organic. Typical procedures used in the production of nanophase or cluster-assembled materials include: (i) creating discrete, smaller clusters that are subsequently fused together in order to produce a bulk-like composite material, or (ii) integrating materials into a compressed and compact (either liquid or solid) matrix composite material throughout the manufacturing process [2, 7, 14].

1.2.1 A brief historical perspective on MNMs

Compared with conventional approaches (filtration and centrifugation), MNMs have the benefits of not using electricity and requiring fewer samples [30, 31]. The majority of the water contaminated by arsenic is located in remote locations with limited power. In these places, the most effective way to clean wastewater is by using MNMs. In a study, nearly 84.9% of the Cd and almost 72% of the Pb were eliminated using a powder composed of Fe_3O_4 NMs and iminodiacetic acid [32]. Since it makes the chelation and ensuing magnetic separation (MS) of heavy metals (HMs) easier, the use of these MNMs is advantageous [33]. BT or NBT and healthcare are two fields in which MNMs are highly useful. The most popular delivery modalities in these applications are internal (*in vivo*) and external (*ex vivo*) administrations. Therapeutic (drug administration as well as hyperthermia) and diagnostic (nuclear magnetic resonance imaging, MRI) usefulness, benefits, and applications are rendered by MNMs. *In vitro* research is mostly focused on diagnostic usefulness, benefits, and applications [34, 35]. Recently, an increasing amount of scientific research has examined MNMs that offer drug delivery capabilities. The development, characterization, and enhancement of MNMs with high-specificity surface functionalization have been the main goals of these investigations. These materials are essential to so many industrial processes and goods in so many diverse industries that modern technology largely depends on them [36].

Numerous distinct features of various magnetic materials allow them to be applied in a multitude of ways. The capacity to change shape allows MNMs to exist in a range of configurations in polymorphic composites or materials. The particle size and the degree or extent of hydration vary, along with other physical attributes such as magnetic character and chemical attributes like the bonds between atoms or ions. Superparamagnetic iron oxide particles are often categorized based on their size, which is less than 10 nm. This is due to the fact that saturation magnetization or full spin pairing is inhibited by monodomain magnetic particles smaller than 10 nm [37].

Iron oxides are MNMs with high biocompatibility that find extensive uses in BT and NBT when compared to other MNMs based on oxides or unalloyed/pure metals. In medical procedures, magnetite (Fe_2O_3) and ferrite (Fe_3O_4) are the two types of iron oxides that are used [38]. High magnetic moments, as well as chemical stability under physiological settings, low toxicity, and a low cost of production, are requirements for MNMs used in BT or NBT. It is noteworthy that the efficacy of MNs in therapeutic, pharmacological, and diagnostic procedures, such as MRI and hyperthermia, is significantly influenced by their dimensions, form, extent of crystallization, and dispersion. They have an obvious connection with these criteria due to the way they are produced. Carbon-powder-coated magnets are currently being investigated for possible uses in numerous industries. These NMs can be coated with carbon nanotubes, which not only bind to a range of other materials but also preserve their magnetic capabilities by adhering to metals, catalysts, polymers, and organic molecules. The nanoparticles are encased in this coating [39].

1.2.2 Property-based classification and development of MNMs

Generally speaking, the chemical composition of MNMs and their fabrication procedure determine their properties. The majority of these properties are the focus of this section.

Magnetic-based properties: MNMs can exhibit superparamagnetism and are characteristically between 1 and 100 nm in size. These particles or composites have no hysteresis or coercivity, since superparamagnetism is induced by thermal processes that are sufficient in strength to spontaneously demagnetize an assembly that has previously been saturated. In this form, the nanoparticles have a far higher magnetic susceptibility and can be magnetized by an external MF. MNMs are nonmagnetic when the field is removed. Targeted drug distribution and controlled therapy may benefit from this feature [40, 41].

Magnetocaloric effect (MCE): The MCE is a phenomenon where some magnetic materials heat up when they are exposed to an MF and cool down when they are removed from it. MNMs' size-dependent superparamagnetic properties make them a viable substitute for traditional bulk materials. Furthermore, MNMs may be able to provide better heat exchange with the environment due to their large surface area. It would be practicable to regulate and monitor the exchange of heat between the MNMs and the surrounding or neighboring matrix by carefully crafting core–shell architectures, opening the door to potential improvements in therapeutic technologies like hyperthermia [41].

1.2.3 Surface properties and charge

According to a hypothesis proposed by Veiseh *et al* [42], in keeping with the biomedical applications of MNMs, proteins absorb charged MNMs and remove them from the bloodstream. Positive MNMs also bind to nonspecific cells, while substantial negative MNMs, in particular, boost liver absorption. When an external electric field is utilized, an electric potential distribution is produced, which establishes the fundamentals of electrophoresis (which has to do with the movement and transmission of dispersed colloids within a fluid). Durán *et al* [43] illustrated the manufacture and use of nanoparticles, including coating and drug loading.

1.2.4 Size dependence

Comparing MNMs to their bulk counterparts reveals a variety of unique magnetic phenomena. Their basic magnetic qualities, such as susceptibility and coercivity (H_c), are affected by their different sizes, shapes, and compositions. When an NM's size drops below a critical value (Dc), it becomes a single magnetic domain; when the temperature increases above the 'blocking temperature,' it displays superparamagnetic activity (Tb). These particular NMs react quickly to applied MFs and exhibit characteristics similar to those of massive paramagnetic atoms with very little coactivity (the field required for the reduction of the magnetization to zero) or remanence (residual magnetism). Superparamagnetic NMs are highly desirable as MR contrast agents because of their characteristics [40, 41].

1.2.5 Composition, shape, and size

This section has addressed the compositional component of MNMs. Today, the three most prevalent forms of MNMs are metals, metal alloys, and metal oxides. The most prevalent NMs are silver, gold, iron, cobalt, nickel, and silver [44]. The majority of metal oxide NMs are composed of ferrites ($CoFe_2O_4$ and $Mn0.6Zn0.4Fe_2O_4$) and iron oxides (Fe_2O_3 and Fe_3O_4); other metal alloy NMs consist of FeCo, FePt, and other elements. As in the case of MNMs, there are several types available now. Many more can be employed for specific purposes, although Fe oxides and Fe with a relatively constant composition are the most common [40, 41].

Currently, ferrites, in combination with other substances such as Zn, Ni, Co, and rare-earth (RE) metals, are deployed in biomedical applications. Owing to the extreme magnetic characteristics that Fe and pure FeCo alloy particles may attain, pure metal particles are gaining attention. A very significant magnetic uniaxial anisotropy is desired in some materials, such as FePt, for utilization in permanent magnets or storage media technologies. Permanent magnets are typically made from RE alloys such as Sm–Co or Nd–Fe–B, but contemporary efforts to create magnets lacking REs have led to the discovery of MNMs composed of Co carbides as well as Fe that have extraordinarily high anisotropy and extremely low iron contents [45].

Size: this parameter, which has no set value, is influenced by the technique used to calculate the particle size. The following radii, or corresponding diameters, can be distinguished:

- Visual radius: transmission electron microscopy (TEM) and atomic force microscopy can be used to determine the physical particle's radius.
- Diffraction radius: this is the size of the coherent diffracting section of the particle. A chaotic surface layer that is virtually noticeable and observable using TEM and has no effect on diffraction may exist on single-crystal particles.
- Magnetic radius: the effective radius of a particle's magnetic core determines how sensitive it is to MFs. Even in well-ordered particles, the magnetic activity varies since the surface atoms have fewer neighbors and, therefore, a reduced number of 'exchange and anisotropy interactions.' This is significant when we consider the superparamagnetic field [40, 41].

1.2.6 Particle shape

Since crystals are transparent and possess ideal crystallographic planes at their surfaces, it is difficult to locate true spherical particles. It is possible to change the kind and concentration of the precursors as well as add impurities during synthesis in an organic medium. At high precursor concentrations, the NM's creation is kinetically controlled, resulting in a cubic or tetrahedral structure. The realm of biomedicine may find a use for nanocapsules and iron oxide nanorods/nano-rice with rod-like structures. Octopus shapes, flower shapes, and other oddities can also be made using original techniques [23].

1.3 Potential applications of MNMs

In view of their numerous potential applications, MNMs are one of the most significant, crucial, and developing categories of materials in BT/NBT. According to Aigbe *et al* [2], Chaudhary and Chaudhary [46], as well as other reported studies, these MNMs find applications in wastewater treatment, sensing, catalysis, MS, magnetic data storage, and a variety of biomedical and environmental fields.

As an illustration, MNMs have been used in MRI as contrast agents. MNMs can also act as potent heat generators in the presence of an alternating MF, destroying tumor cells [46]. This property makes the use of these NMs in cancer hyperthermia therapy possible. Drug delivery agents that may be localized within the body at a desired location with the application of an external MF have also been employed in conjunction with MNMs [46].

1.3.1 Magnetic separation

Most subfields of BT/NBT require the isolation and separation of specific particles or molecules, such as DNA, proteins, and cells, before beginning any biomedical and pharmacological research. As a result of its distinctive MS properties and prospective efficiency, MNM-based bioseparation is one of the most studied and frequently applied bioseparation techniques. During this procedure, MNM colloids are used to identify the biological molecules. An external MF is then used to separate the molecules. This process can be useful for RNA/DNA extraction, cell isolation, protein purification, and immunoprecipitation. The tiny size, potential separation performance, and strong dispersibility of MNM particles, such as beads, have made them popular for cell and biomolecule separation and purification. Using antibodies coupled with beads to produce extremely precise antibodies that can specifically bind to their matching antigens on the surface of the targeted locations is one of the most recent developments in the MS field of study [46].

1.3.2 Diagnostics

The use of MNMs to label stem cells has led to the growth, development, and advancement of noninvasive imaging procedures. Among these, MRI is a commonly used diagnostic technology that provides excellent anatomical detail and high spatial resolution for visualizing the structure and function of tissues. Improved sensitivity, superior biocompatibility, and quick detection at modest doses are only a few of the major advantages of the various types of MNMs that have been produced to improve contrast agents in MRI imaging [46].

1.3.3 Sensors

Given that MNMs have special magnetic properties not seen in biological systems, many types of MNM-based BSs/NBSs have been surface functionalized to recognize particular chemical targets. MNMs can be employed for biosensing/nanobiosensing, offering an improvement in sensitivity and stability in a range of devices and formats due to their distinct composition, size, and magnetic properties [46, 47].

1.3.4 Drug delivery

The development of MNMs has enabled tumor-specific, localized medication delivery to be achieved. Initially, the medicine is carried by an MNM and is either dissolved in the coating or adhered to its exterior. A powerful permanent magnet creates an MF gradient to hold the drug-coated particles in the desired area once they have been added to the patient's bloodstream. The use of MNMs coated with medicine is an extremely promising drug delivery technique, since they may be transported, held in specific places, and injected continuously [46, 47].

1.3.5 Therapy

Currently under investigation, MNMs may be used to achieve therapeutic heating of tumors, often known as hyperthermia. For particular tumor areas, several kinds of superparamagnetic NMs with distinct coatings and targeting agents are employed. The depths required to treat cancer almost anywhere in the human body can be reached by magnetic particle heating. Furthermore, MNM hyperthermia exhibits considerable promise as an adjunct to radiation therapy and conventional chemotherapy [46, 47].

1.4 Review of research on the potential applications of MNMs

1.4.1 Environmental domains

As already stated, MNMs have several benefits and applications in BT/NBT and environmental restoration, remediation, and management [5, 41]. They may be used to extract large amounts of organic and inorganic compounds from environmental matrices [23, 48–51]. These complexes contain a variety of chemicals, including HMs, pigments, oils, medicines, pesticides, and nutritional elements [19, 23]. The MNMs' physical, chemical, and pharmacological features [52, 53] have to satisfy stringent requirements for BT/NBT applications in terms of their chemical composition, granulometric uniformity, crystal structure, magnetic behavior, surface structure, adsorption properties, solubility, and low self-toxicity, as reported by Mohammed *et al* [54]. MNMs and contaminating molecules may react by coprecipitation, intense reductive reactivity, or electrostatic attraction. In order to increase the effectiveness of contaminant removal, MNMs can also be employed in water management and purification as non-absorbents or immobilization carriers [44, 51]. These applications and benefits offer substantial environmental advantages, such as substantial removal efficiency (which is typically higher than 85%), faster kinetics (that is, first- and second-order kinetics), and more powerful reactivity with numerous contaminants (both organic and inorganic) due to their small size at the nanoscale, relatively large surface area (10–500 m^2 g^{-1}), superparamagnetic behavior, and biocompatibility with biomolecules [55]. MNMs are a crucial tool in the selective and extremely efficient adsorption of hazardous HM ions from various matrices, such as Pb^{2+}, Cd^{2+}, Cu^{2+}, Zn^{2+}, Ni^{2+}, and As^{3+} [23, 56, 57]. Diethylenetriamine-modified chitosan MNMs have an excellent neutral pH adsorption capability and are resistant to acid [58]. They are capable of eliminating ions of rare-earth metals such as Dy^{3+}, Nd^{3+}, and Er^{3+}. Moreover, radioactive metal ions may be effectively eliminated by MNMs. With an efficiency of around 96%, radioactive uranium (U) could possibly be extracted from an aqueous solution using MNMs of CeO_2–TiO_2–Fe_2O_3 [59]. Another important application for MNMs is the removal of dyes/color from industrial waste/effluent, such as that produced by the painting, tannery, dyeing, and textile sectors [60–63]. It has been reported that bare MNMs, such as magnetite, ferrite, and maghemite NMs, can be used as innovative adsorbents for the separation and management of dyes. Anionic as well as cationic dyes can be eliminated by maghemite [60–64] and magnetic iron oxide nanoparticles (MIONs) [65]. Magnetic core–shell NMs and surface-coated MNMs can be used to eliminate dyes. To extract Acid Black 1 from various aqueous environments, Fe_3O_4 has been used to electrochemically synthesize an innovative magnetic adsorbent known as an Fe_3O_4/Al_2O_3 hybrid composite or compound [66]. An anionic sodium dodecyl sulfate (SDS) surfactant has been used to modify $ZnFe_2O_4$ NMs, which were electrostatically attracted to cationic methylene blue. At pH = 12, an adsorption capacity of about 699.30 mg g^{-1} was attained [67].

1.4.2 Biomedical domains

The permanence and hydrophobic properties of oil in soils and groundwater pose serious risks to human well-being and the environment [1, 68–72]. Over 85% of the oil in wastewater may be removed by MNMs with a surface coating. MNMs with three distinct filtration layers (non-porous silica as the interlayer, mesoporous silica as the outer layer, and ferro–ferric oxide as the core) are used to clean oilfield wastewater [73]. Recently, polyoxyalkylated N-dimethylethanolamine–Fe3O4 MNMs have been employed to rapidly separate emulsified oil droplets and achieve high oil removal efficacy (92.30%) [74]. Hydrophobic MNMs coated on the surface have a strong affinity for oil and may effectively extract oils with varying viscosities from the water's surface, irrespective of their density [75, 76]. Over the last few decades, pharmaceuticals have been a concern for soils, vadose zones, and groundwater [77, 78]. Some studies have focused on the synthesis of MNMs for the purpose of removing pharmaceutical products from environmental media. With absorption capabilities of 25.00 mg g^{-1} at pH 4.00 and 30.40 mg g^{-1} at pH 6.00, nickel ferrite NMs are a potential dipyrone adsorbent [79]. Chitosan–Fe_3O_4 MNMs have the ability to adsorb tetracycline, and this adsorption exhibits pseudo-second-order (PSO) kinetics [63]. Amoxicillin can be eliminated as an adsorbent by employing magnetic Fe_3O_4@C NMs, in accordance with the PSO kinetic models and Freundlich and Langmuir isotherms [80]. Consequently, some MNM research has focused on pesticide remediation because pesticides are highly persistent in the environment and take a lengthy period to naturally attenuate in soils. This poses a serious risk to human health as well as the health of ecosystems [81]. According to recent research, surface imprinting technology and the sol–gel process can be utilized in removing di-(2-ethylhexyl) phthalate from water by employing polymer NMs with molecularly imprinted patterns [82]. For graphene oxide/MION NMs, the removal/confiscation efficacy of isomeric endrin and dieldrin pesticides is 86.0% and 74.0%, respectively, exhibiting distinct stereoselective removal performance [83]. It has been demonstrated that MNMs treated with ionic liquids have increased extraction efficiency for a number of pesticides, including parathion, fenthion, phoxim, temephos, and paraquat [84]. Recently, MNMs have been used for the adsorption and recovery of nutrients, such as phosphates and nitrates, from industrial wastewater facilities. For instance, Al-doped MNMs have effectively removed phosphates from urban wastewater and chicken washing effluent [85]. Adsorbents consisting of MNMs of ($MgFe_2O_4$) and biochar may be able to recover phosphate from wastewater/effluent. They are recyclable and incredibly effective [66]. The adsorption efficacy of magnetically activated carbons for nitrate ions is 76.29% at pH = 4, with an equilibrium duration of 1200 s [86]. Even more exciting are the numerous environmental cleanup systems that can be built with MNMs [87]. With improved comprehension and manipulation of nanomagnetic phenomena, a plethora of industrial applications, from bank checks to engine seals, are now feasible. We now have a greater understanding and control of nanomagnetic processes [88]. This technology has led to a number of beneficial inventions, including strong permanent magnets and magnetic storage devices that can hold vast amounts of data. Depending on the application, surface effects at the nanoscale may be advantageous or disadvantageous [89].

Given that even the smallest 'bit' of information is susceptible to being limited and because such structures are probably bound by magnetic interactions, the magnetic storage industry has given MNMs a lot of attention [90, 91]. The most recent advancements in the study of MNMs and their applications to BT/NBT and environmental remediation are discussed in this book.

1.5 Conclusions and future directions

The scientific community is very interested in the various applications of MNMs because of their unique chemical and physical characteristics that set them apart from other adsorbent materials and greatly improve food, environmental, and health safety. Given that they are easily separated and recovered by an external magnet and may be reused multiple times, they are frequently utilized in these operations.

As this chapter has highlighted, MNMs can be readily modified to increase their surface area by adding different materials, such as graphene, ionic liquids, carbon nanotubes, and polymers, among others. This improves their extraction efficacy and reduces the trace levels of the analytes targeted within complicated matrices. Due to their unique extraction capacities, which also allow MNMs to be easily modified and have a large surface area and charge transfer capacity, they are used in a wide range of extraction applications. These MNMs and their NMs have demonstrated a great deal of promise for the extraction of different pollutants, and they may find use in the effective analysis of other significant chemical contaminants found in food, human bodies, and the environment, as well as in new pollutants.

There are still gaps in the application of some alloys (FeCo, FePt, and FePd) for pollution extraction, despite the fact that MNMs and their alloys have been extensively used in the fields of medical research, microwave absorption, catalysis, and magnetic data storage/magnetic recording media, in contrast to BT/NBT. As a result, there are gaps in our understanding of how well MNMs may be used to extract environmental toxins, especially new (emerging) pollutants. Additional areas of interest might be the synthesis and/or modification of MNMs using contemporary synthetic methods, such as naturally generated plant extracts and environmentally friendly solvents such as ethanol and water.

References

[1] Ukhurebor K, Aigbe U, Onyancha R, Athar H, Okundaye B, Aidonojie P, Siloko B, Hossain I, Kusuma H and Darmokoesoemo H 2024 Environmental influence of gas flaring: perspective from the Niger Delta Region of Nigeria *Geofluids* **2024** 1–17

[2] Aigbe U, Ukhurebor K and Onyancha R 2023 *Magnetic Nanomaterials: Synthesis, Characterization and Applications* (Berlin: Springer Nature)

[3] Singh R, Ukhurebor K, Singh J, Adetunji C and Singh K 2022 *Nanobiosensors for Environmental Monitoring: Fundamentals and Application* (Berlin: Springer Nature)

[4] Kerry R *et al* 2021 A comprehensive review on the applications of nano-biosensor based approaches for non-communicable and communicable disease detection *Biomaterials* **9** 3576–602

[5] Bushra R, Ahmad M, Alam K, Farzad S, Qurtulen, Shakeel S *et al* 2024 Recent advances in magnetic nanoparticles: key applications, environmental insights, and future strategies *Sustain. Mater. Technol.* **40** e00985

[6] Khan I, Saeed K and Khan I 2019 Nanoparticles: properties, applications and toxicities *Arab. J. Chem.* **12** 908–31

[7] Ukhurebor K and Aigbe U 2024 *Environmental Applications of Magnetic Sorbents* (Bristol: Institute of Physics (IOP) Publishing)

[8] Ukhurebor K, Aigbe U and Onyancha R 2023 *Adsorption Applications for Environmental Sustainability* (Bristol: Institute of Physics (IOP) Publishing)

[9] Jeevanandam J, Barhoum A, Chan Y, Dufresne A and Danquah K 2018 Review on nanoparticles and nanostructured materials: history, sources, toxicity and regulations *Beilstein J. Nanotechnol.* **9** 1050–74

[10] Sen M 2020 *Nanotechnology and the Environment* (London: IntechOpen)

[11] Baig N, Kammakakam I and Falath W 2021 Nanomaterials: a review of synthesis methods, properties, recent progress, and challenges *Mater. Adv.* **2** 1821–71

[12] Basavegowda N and Baek K-H 2021 Multimetallic nanoparticles as alternative antimicrobial agents: challenges and perspectives *Molecules* **26** 912

[13] Alsaiari N, Katubi K, Alzahrani F, Siddeeg S and Tahoon M 2021 The application of nanomaterials for the electrochemical detection of antibiotics: a review *Micromachines* **12** 308

[14] Baby R, Hussein M, Abdullah A and Zainal Z 2022 Nanomaterials for the treatment of heavy metal contaminated water *Polymers* **14** 583

[15] Carneiro P, Morais S and Pereira M 2019 Nanomaterials towards biosensing of Alzheimer's disease biomarkers *Nanomaterials* **9** 1663

[16] Cheng Z, Li M, Dey R and Chen Y 2021 Nanomaterials for cancer therapy: current progress and perspectives *J. Hematol. Oncol.* **14** 85

[17] Huang C, Lin C, Nguyen M, Hussain A, Bui X and Ngo H 2023 A review of biosensor for environmental monitoring: principle, application, and corresponding achievement of sustainable development goals *Bioengineered.* **1** 58–80

[18] Lu F and Astruc D 2018 Nanomaterials for removal of toxic elements from water *Coord. Chem. Rev.* **356** 147–64

[19] Malakar A, Kanel S, Ray C, Snow D and Nadagouda M 2021 Nanomaterials in the environment, human exposure pathway, and health effects: a review *Sci. Total Environ.* **759** 143470

[20] Maleki Dizaj S, Eftekhari A, Mammadova S, Ahmadian E, Ardalan M, Davaran S, Nasibova A, Khalilov R, Valiyeva M, Mehraliyeva S *et al* 2021 Nanomaterials for chronic kidney disease detection *Appl. Sci.* **11** 9656

[21] Yang J, Hou B, Wang J, Tian B, Bi J, Wang N, Li X and Huang X 2019 Nanomaterials for the removal of heavy metals from wastewater *Nanomaterials* **9** 424

[22] Abdel Maksoud M, Fahim R and Shalan A E A 2021 Advanced materials and technologies for supercapacitors used in energy conversion and storage: a review *Environ. Chem. Lett.* **19** 375–439

[23] Ali A, Shah T, Ullah R, Zhou P, Guo M, Ovais M, Tan Z and Rui Y 2021 Review on recent progress in magnetic nanoparticles: synthesis, characterization, and diverse applications *Front. Chem.* **9** 629054

[24] Soto-Bernal J, Gonzalez-Mota R, Rosales-Candelas I and Ortiz-Lozano J 2015 Effects of static magnetic fields on the physical, mechanical, and microstructural properties of cement pastes *Adv. Mater. Sci. Eng.* **2015** 1–9

[25] Chen X, Charrier M and Srubar W 2021 Nanoscale construction biotechnology for cementitious materials: a prospectus *Front. Mater.* **7** 594989

[26] Bayda S, Adeel M, Tuccinardi T, Cordani M and Rizzolio F 2019 The history of nanoscience and nanotechnology: from chemical-physical applications to nanomedicine *Molecules* **25** 112

[27] Barhoum A, García-Betancourt M, Jeevanandam J, Hussien E, Mekkawy S, Mostafa M, Omran M, Abdalla M and Bechelany M 2022 Review on natural, incidental, bioinspired, and engineered nanomaterials: history, definitions, classifications, synthesis, properties, market, toxicities, risks, and regulations *Nanomaterials* **12** 177

[28] Granqvist C and Buhrman R 1976 Ultrafine metal particles *J. Appl. Phys.* **47** 2200–19

[29] E. Commission 2012 Nanotechnologies: principles, applications, implications and hands on activities (a compendium for educators) *LU: Directorate General for, Reasearch and Innovation* (Luxembourg: Publications Office of the European Commission)

[30] Liosis C, Papadopoulou A, Karvelas E, Karakasidis T and Sarris I 2021 Heavy metal adsorption using magnetic nanoparticles for water purification: a critical review *Materials (Basel)* **14** 7500

[31] Nisticò R, Celi L, Bianco Prevot A, Carlos L, Magnacca G, Zanzo E and Martin M 2018 Sustainable magnet-responsive nanomaterials for the removal of arsenic from contaminated water *J. Hazard. Mater.* **342** 260–9

[32] Morales-Amaya C, Alarcón-Herrera M, Astudillo-Sánchez P, Lozano-Morales S, Licea-Jiménez L and Reynoso-Cuevas L 2021 Ferrous magnetic nanoparticles for arsenic removal from groundwater *Water* **13** 2511

[33] Fan L, Song J, Bai W *et al* 2016 Chelating capture and magnetic removal of non-magnetic heavy metal substances from soil *Sci. Rep.* **6** 21027

[34] Anik M, Hossain M, Hossain I, Mahfuz A, Rahman M and Ahmed I 2021 Recent progress of magnetic nanoparticles in biomedical applications: a review *Nano Select.* **2** 1146–86

[35] Baghban R, Afarid M, Soleymani J and Rahimi M 2021 Were magnetic materials useful in cancer therapy? *Biomed. Pharmacother.* **144** 112321

[36] Stiufiuc G and Stiufiuc R 2024 Magnetic nanoparticles: synthesis, characterization, and their use in biomedical field *Appl. Sci.* **14** 1623

[37] Ali A, Zafar H, Zia M, Haq I, Phull A, Ali J and Hussain A 2016 Synthesis, characterization, applications, and challenges of iron oxide nanoparticles *Nanotechnol. Sci. Appl.* **9** 49–667

[38] Sangaiya P and Jayaprakash R 2018 A review on iron oxide nanoparticles and their biomedical applications *J. Supercond. Novel Magn.* **31** 3397–413

[39] Mourdikoudis S, Pallares R and Thanh N 2018 Characterization techniques for nanoparticles: comparison and complementarity upon studying nanoparticle properties *Nanoscale* **10** 12871–934

[40] Sahoo H and Sahoo J 2024 *Iron Oxide-Based Nanocomposites and Nanoenzymes: Fundamentals and Applications* (Cham: Springer Nature)

[41] Woldeamanuel M, Mohapatra S, Senapati S, Bastia T, Panda A and Prasanta Rath P 2024 Role of magnetic nanomaterials in environmental remediation *Iron Oxide-Based Nanocomposites and Nanoenzymes: Fundamentals and Applications* (Cham: Springer Nature) 185–208

[42] Veiseh O *et al* 2015 Size and shape-dependent foreign body immune response to materials implanted in rodent and non-human promates *Nat. Mater.* **14** 643–51

[43] Durán A, Kuiper J, Aguiar A *et al* 2023 Bringing the Nature Futures Framework to life: creating a set of illustrative narratives of nature futures *Sustain. Sci.* **2023**

[44] Yang J, Hou B, Wang J, Tian B, Bi J, Wang N, Li X and Huang X 2019 Nanomaterials for the removal of heavy metals from wastewater *Nanomaterials (Basel)* **9** 424

[45] Abdel Maksoud M, Fahim R, Shalan A *et al* 2021 Advanced materials and technologies for supercapacitors used in energy conversion and storage: a review *Environ. Chem. Lett.* **19** 375–439

[46] Chaudhary V and Chaudhary R 2018 Magnetic nanoparticles: synthesis, functionalization, and applications *Encyclopaedia of Nanoscience and Nanotechnology* (New York: American Scientific Publishers) 153–83

[47] Kerry R *et al* 2021 A comprehensive review on the applications of nano-biosensor based approaches for non-communic *Biomater. Sci.* **9** 3576–602

[48] Guo T *et al* 2018 The recent advances of magnetic nanoparticles in medicine *J. Nanomater.* **7805147** 1–8

[49] Mutanda T, Naidoo D, Bwapwa J and Anandraj A 2020 Biotechnological applications of microalgal oleaginous compounds: current trends on microalgal bioprocessing of products *Front. Energy Res.* **8** 598803

[50] Tahir M, Sohaib M, Sagir M and Rafique M 2022 Role of nanotechnology in photocatalysis *Encyclopedia of Smart Materials* (Amsterdam: Elsevier) **2** 578–89

[51] Wadhawan S, Jain A, Nayyar J and Mehta S 2020 Role of nanomaterials as adsorbents in heavy metal ion removal from waste water: a review *J. Water Process Eng.* **33** 101038

[52] Hashem A, Hossain M, Marlinda R, Al Mamun M, Simarani K and Johan M R 2021 Nanomaterials based electrochemical nucleic acid biosensors for environmental monitoring: a review *Appl. Surf. Sci. Adv.* **4** 100064

[53] Nisticò R, Cesano F and Garello F 2020 Magnetic materials and systems: domain structure visualization and other characterization techniques for the application in the materials science and biomedicine *Inorganics* **8** 6

[54] Mohammed L, Gomaa H, Ragab D and Zhu J 2017 Magnetic nanoparticles for environmental and biomedical applications: a review *Particuology* **30** 1–14

[55] Tang S and Lo I 2013 Magnetic nanoparticles: essential factors for sustainable environmental applications *Water Res.* **47** 2613–32

[56] Ge F, Li M, Ye H and Zhao B 2012 Effective removal of heavy metal ions Cd^{2+}, Zn^{2+}, Pb^{2+}, Cu^{2+} from aqueous solution by polymer-modified magnetic nanoparticles *J. Hazard. Mater.* **211–212** 366–72

[57] Hasanzadeh R, Moghadam P, Bahri-Laleh N and Sillanpää M 2017 Effective removal of toxic metal ions from aqueous solutions: 2-bifunctional magnetic nanocomposite base on novel reactive PGMA-MAn copolymer@Fe_3O_4 nanoparticles *J. Colloid Interface Sci.* **490** 727–46

[58] Liu E, Zheng X, Xu X, Zhang F, Liu E, Wang Y, Li C and Yan Y 2017 Preparation of diethylenetriamine-modified magnetic chitosan nanoparticles for adsorption of rare-earth metal ions *New J. Chem.* **41** 7739–50

[59] El-sherif R, Lasheen T and Jebril E 2017 Fabrication and characterization of CeO_2–TiO_2–Fe_2O_3 magnetic nanoparticles for rapid removal of uranium ions from industrial waste solutions *J. Mol. Liq.* **241** 260–9

[60] Bayramoglu G, Altintas B and Arica M 2009 Adsorption kinetics and thermodynamic parameters of cationic dyes from aqueous solutions by using a new strong cation-exchange resin *Chem. Eng. J.* **152** 339–46

[61] Eleryan A, Hassaan M, Aigbe U, Ukhurebor K, Onyancha R, Kusuma H, El-Nemr M, Ragab S and El Nemr A 2023 Biochar-C-TETA as a superior adsorbent to acid yellow 17 dye from water: isothermal and kinetic studies *J. Chem. Technol. Biotechnol.* **98** 2415–28

[62] Eleryan A, Aigbe U, Ukhurebor K, Onyancha R, Hassaan M, Elkatory M, Ragab S, Osibote O, Kusuma H and El Nemr A 2023 Adsorption of direct blue 106 dye using zinc oxide nanoparticles prepared via green synthesis technique *Environ. Sci. Pollut. Res.* **30** 6966–69882

[63] Aigbe U, Ukhurebor K, Onyancha R, Okundaye B, Pal K, Osibote O, Esiekpe E, Kusuma H and Darmokoesoemo H 2022 A Facile Review on the Sorption of Heavy Metals and Dyes using Bionanocomposites *Adsorpt. Sci. Technol.* **8030175** 1–36

[64] Afkhami A and Moosavi R 2010 Adsorptive removal of Congo red, a carcinogenic textile dye, from aqueous solutions by maghemite nanoparticles *J. Hazard. Mater.* **174** 398–403

[65] Qadri S, Ganoe A and Haik Y 2009 Removal and recovery of acridine orange from solutions by use of magnetic nanoparticles *J. Hazard. Mater.* **169** 318–23

[66] Jung K-W, Choi B, Ahn K-H and Lee S-H 2017 Synthesis of a novel magnetic Fe_3O_4/γ-Al_2O_3 hybrid composite using electrode-alternation technique for the removal of an azo dye *Appl. Surf. Sci.* **423** 383–93

[67] Zhang R, Wang Z, Zhou Z, Li D, Wang T, Su P and Yang Y 2019 Highly effective removal of pharmaceutical compounds from aqueous solution by magnetic Zr-based MOFs composites *Ind. Eng. Chem. Res.* **58** 3876–84

[68] Ukhurebor K, Ngonso B, Egielewa P, Cirella G, Akinsehinde B and Balogun V 2023 Petroleum spills and the communicative response from petroleum agencies and companies: impact assessment from the Niger Delta Region of Nigeria *Extract. Indust. Soc.* **15** 101331

[69] Ukhurebor K, Hussain A, Adetunji C, Aigbe U, Onyancha R and Abifarin O 2021 Environmental implications of petroleum spillages in the Niger Delta Region of Nigeria: a review *J. Environ. Manage.* **293** 112872

[70] Emegha J, Oliomogbe T, Okpoghono J, Babalola A, Ejelonu C, Elete D and Ukhurebor K 2023 Green biosorbents for the degradation of petroleum contaminants *Adsorption Applications for Environmental Sustainability* (Bristol: Institute of Physics Publishing) 12

[71] Anani A, Adama K, Ukhurebor K, Habib A, Abanihi V and Pal K 2023 Application of nanofibrous protein for the purification of contaminated water as a next generational sorption technology: a review *Nanotechnology* **34** 1–18

[72] Alegbeleye O, Opeolu B and Jackson V 2017 Polycyclic aromatic hydrocarbons: a critical review of environmental occurrence and bioremediation *Environ. Manage* **60** 758–78

[73] Liu Z, Yang H, Zang H, Huang C and Li L 2012 Oil-field wastewater purification by magnetic separation technique using a novel magnetic nanoparticle *Cryogenics* **52** 699–703

[74] Duan M, Xu Z, Zhang Y, Fang S, Song X and Xiong Y 2017 Core–shell composite nanoparticles with magnetic and temperature dual stimuli-responsive properties for removing emulsified oil *Adv. Powder Technol.* **28** 1291–7

[75] Dutra G, Araújo O, Neto W, Garg V, Oliveira A and J A F 2017 Obtaining superhydrophopic magnetic nanoparticles applicable in the removal of oils on aqueous surface *Mater. Chem. Phys.* **200** 204–16

[76] Shahzad A, Aslibeiki B, Slimani S, Ghosh S, Vocciante M, Grotti M, Comite A, Peddis D and Sarkar T 2024 Magnetic nanocomposite for lead (II) removal from water *Sci. Rep.* **4** 17674

[77] Gomes A, Justino C, Rocha-Santos T, Freitas A, Duarte A and Pereira R 2017 Review of the ecotoxicological effects of emerging contaminants to soil biota *J. Environ. Sci. Health A Tox. Hazard Subst. Environ. Eng.* **52** 992–1007

[78] Ma L, Liu Y, Zhang J, Yang Q, Li G and Zhang D 2018 Impacts of irrigation water sources and geochemical conditions on vertical distribution of pharmaceutical and personal care products (PPCPs) in the vadose zone soils *Sci. Total Environ.* **626** 1148–56

[79] Springer V, Pecini E and Avena M 2016 Magnetic nickel ferrite nanoparticles for removal of dipyrone from aqueous solutions *J. Environ. Chem. Eng.* **4** 3882–90

[80] Kakavandi B, Esrafili A, Mohseni-Bandpi A, Jonidi Jafari A and Kalantary R 2014 Magnetic Fe_3O_4@C nanoparticles as adsorbents for removal of amoxicillin from aqueous solution *Water Sci. Technol.* **69** 147–55

[81] Odukkathil G and Vasudevan N 2013 Toxicity and bioremediation of pesticides in agricultural soil *Rev. Environ. Sci. Biotechnol.* **12** 421–44

[82] Li C, Ma X, Zhang X, Wang R, Li X and L Q 2017 Preparation of magnetic molecularly imprinted polymer nanoparticles by surface imprinting by a sol-gel process for the selective and rapid removal of di-(2-ethylhexyl) phthalate from aqueous solution *J. Sep. Sci.* **40** 1621–8

[83] Shrivas K, Ghosale A, Nirmalkar N, Srivastava A, Singh S and Shinde S 2017 Removal of endrin and dieldrin isomeric pesticides through stereoselective adsorption behavior on the graphene oxide-magnetic nanoparticles *Environ. Sci. Pollut. Res. Int.* **24** 24980–8

[84] Latifeh F, Yamini Y and Seidi S 2016 Ionic liquid-modified silica-coated magnetic nano-particles: promising adsorbents for ultra-fast extraction of paraquat from aqueous solution *Environ. Sci. Pollut. Res.* **23** 4411–21

[85] Xu J, Luu L and Tang Y 2017 Phosphate removal using aluminum-doped magnetic nanoparticles *Desalin. Water Treat.* **58** 238–49

[86] Arbabi M, Hemati S, Shamsizadeh Z and Arbabi A 2017 Nitrate removal from aqueous solution by almond shells activated with magnetic nanoparticles *Desalin. Water Treat.* **80** 344–51

[87] Patra J *et al* 2018 Nano based drug delivery systems: recent developments and future prospects *J. Nanobiotechnol.* **16** 71

[88] Samrot A V, Sahithya C S, Selvarani A J, Purayil S K and Ponnaiah P 2021 A review on synthesis, characterization and potential biological applications of superparamagnetic iron oxide nanoparticles *Curr. Res. Green Sustain. Chem.* **4** 100042

[89] Baig N, Kammakakam I and Falath W 2021 Nanomaterials: a review of synthesis methods, properties, recent progress, and challenges *Mater Adv.* **2** 1821–71

[90] Peddis D, Jönsson P E, Laureti S and Varvaro G 2014 Magnetic interactions: a tool to modify the magnetic properties of materials based on nanoparticles *Nanomagnetism: Fundamentals and Applications* (Frontiers of Nanoscience vol 6) (Amsterdam: Elsevier) pp 129–88

[91] Wallyn J, Anton N and Vandamme T F 2019 Synthesis, principles, and properties of magnetite nanoparticles for *in vivo* imaging applications—a review *Pharmaceutics* **11** 601

IOP Publishing

Environmental Applications of Magnetic Nanomaterials
Under the Influence of Magnetic Fields

Uyiosa Osagie Aigbe and Kingsley Eghonghon Ukhurebor

Chapter 2

Fabrication and characterization of magnetic nanomaterials and their application in contaminant removal under the effect of a magnetic field

Organic compound contamination and accumulation in the ecosystem is a considerable and universal problem owing to swift industrialization and the creation and continuous discharge of industrial contaminants into the ecosystem. There has been a global trend toward the growing use of novel technologies such as magnetic field (MF)-assisted adsorption (ADPN) processes using magnetic nanoparticles (MNPs) in the remediation of industrial effluents. For acceptable results, the intensity and resolution of the MF and the magnetic properties of the sorbent and sorbate are pivotal. This alternative contaminant treatment process is important in today's world due to its ease of use, ecological purity, and safety.

2.1 Introduction

Water is an essential source for human survival. According to the World Water Development Report released by the United Nations Educational, Scientific and Cultural Organization (UNESCO) in 2021, international freshwater usage increased sixfold in the last century and has been increasing by 1% per annum since the 1980s. As water consumption increases, the quality of water faces severe challenges [1]. Thus, the supervision and regulation of industrial runoff are country-specific, and various nations have diminutive or no actual legal agendas supported by regulatory frameworks. According to the UNESCO report, in developing countries, 70%–80% of industrial runoffs are dumped untreated. Yet, as most nations are tightening their

doi:10.1088/978-0-7503-6377-8ch2

regulatory agendas, industries are confronted with meeting stricter requirements for water release and reuse [2, 3].

Organic compound contamination and accumulation in the ecosystem is a considerable universal problem owing to swift industrialization and the creation and continuous discharge of industrial contaminants into the ecosystem. Heavy metals (HMs), dyes, and other chemical contaminants present in unprocessed effluents have negative impacts on the health of humans [4]. Mining expansion, smelting, battery production, chemical/biochemical industries, and the extensive use of chemical/biochemical substances for agricultural activities have released vast quantities of HMs and dyes into surface water and groundwater. Organic contaminants, inorganic (INORG) contaminants, and microorganisms are the three key forms of contaminants found in water. HM ions (HMIs) are a major source of inert pollutants due to their environmental toxicity and biotic activity. As a result, the global community has shown increasing concern about these HMIs in recent years [5].

Another significant source of organic industrial contaminants is dyes. They are generally employed in paint, textile, pigment, paper, and wool production. They cannot be removed during the process of waste release, resulting in colored water [4]. Also, the inappropriate and unsafe usage of drugs and specific agrochemicals has triggered the pollution of watercourses and ultimately the food chain. Antibiotic contamination also leads to the rise of superbugs or multi-drug-resistant microbial species, increasing the challenge of creating advanced novel antibiotics. Antibiotics have been shown to harm the fauna and flora of the receiving water body. They not only influence the exquisite balance of the ecosystem but also stop sunlight from penetrating the water body's surface, hence decreasing photosynthesis in marine flora. Thus, it is essential to remove these contaminants before they are discharged into water bodies [6].

Today, experts, governments, and water regulation agencies are concerned about sustaining and improving water quality. Thus, novel technologies that can reutilize industrially produced effluents are in demand. The treatment/remediation of effluents is a key global concern, especially when key features such as sustainability, cost, recyclability, and reusability are taken into view [5, 7, 8]. Traditional water treatment (WT) approaches are unable to sequester all INORG and organic contaminants and pathogens from water, and employing a single traditional approach to remove various contaminants is not viable [4].

A substantial growth area in WT is the innovative quality of modest, low-cost, and enhanced technology. With an increased demand for water combined with mounting global energy costs, more energy cannot be expended by industries for the treatment of wastewater (WWR). Hence, the requirement for well-organized water remediation approaches that offer substantial benefits in terms of both manufacturing time and cost cannot be overstated. Approaches such as chemical oxidation, ion exchange, filtration, membrane separation, flocculation, ADPN, oxidation–reduction, and electrochemical methods have been utilized in the confiscation of dissolved contaminated substances such as INORG and organic contaminants [9].

However, most of these approaches lead to incomplete contaminant confiscation, elevated amounts of reagents and energy expenditure, sludge generation, and a requirement for dedicated apparatus or monitoring systems for water remediation. Sorption is well regarded as an efficient and economical approach to water remediation owing to its straightforward design and ease of handling [9, 10]. The sorption method can be significantly improved by employing external features such as ultrasonic waves, MFs, radiation, and electric fields (EFs). In WT, the use of MFs is valued due to their easy access, effectiveness, performance, economical energy consumption, and minimal influence on the ecosystem [11].

MNPs' key benefits are their substantial pollutant sequestration capability and swift reaction rates. A particle's total surface area (SA) intensifies enormously (for the same mass) when the particle size is reduced to the nanoscale (NS). Owing to the extraordinary ratio of SA to volume, there is an increase in the active sites for the reaction, and the required mass for the treatment processes can be smaller than the mass required for the corresponding micron-sized items [12]. Though conventional WWR treatment (WWT) approaches have been successfully used in an extensive range of WWR scenarios, they are far from effective in the treatment of industrial WWR that contains large volumes of HMs, dyes, or noxious substances. Under these conditions, the use of MFs improves effluent treatment approaches owing to their small energy consumption, environmental friendliness, and extraordinary effectiveness [13].

A distinct consideration is the magnetic treatment of WWR employing physical and chemical approaches, owing to their environmental cleanliness, simplicity, and safety. This emerging novel procedure for the remediation of water has attracted the attention of researchers worldwide. Generally, the technical systems are coupled with specific physical characteristics. In the sorption process (SP), variations in the sorbent or sorbate response to the MF depend on the field strength, the applied field direction, the MF exposure time, the rate of solution flow, the additives present in the system and the pH of the medium [9–11].

Different sectors of society have been transformed by diverse nanoparticle (NP) applications. NPs are tiny particles that occur in a typical size range of 1–100 nm, which differentiates them from their bulk parent materials and makes them ideal materials for various applications. Among them are MNPs, NS materials with unique magnetic features that have broad applications in various fields. In recent years, they have gained substantial attention due to their use in specific sectors such as biosensing, medicine, agriculture, the environment, cancer theranostics, and catalysis. It is critical to design MNPs using controlled surface engineering to meet the requirements of their intended applications. Recently, the progress seen in the field of nanotechnology (NT) has contributed to the advancement and transformation of various sectors [14].

MNPs are created from various metal elements, either alone or in composites, and their oxides with magnetic features. MNPs are a distinct kind of NP that can interact with an external MF (EMF) as an inevitable consequence of their ferrimagnetic, ferromagnetic, and superparamagnetic features. Specifically, superparamagnetic magnetite (Fe_3O_4) is the most frequently employed type of iron oxide (FeO) owing to its extraordinary biocompatibility and low toxicity. Recently, FeO MNPs have undergone significant development and have been considered for use in various areas. Features such as metal-rich moieties, extraordinary SA, and tunable structures make MNPs a field of great interest, as MNPs may find extensive use in biomedical applications, drug delivery, catalysis, bioimaging, and environmental applications. In the last decade, their tunable shape and size properties have made them a hot topic [14, 15]. A characteristic MNP architecture comprises a coating and a core–shell structure, with the core generally including magnetic elements such as nickel (Ni), Fe, cobalt (Co), etc. and their respective oxides. The covering shell is responsible for protecting the NPs from the medium, thus conserving their properties and structure. MNPs' extraordinary SA permits easy modification of their surface, making them versatile for different uses [15]. The use of MFs in the water purification/WWT process and the mechanism of this WWT approach, namely the use of MNPs, are presented in this book chapter, as the presence of noxious HMs in surface waters and effluents signifies a severe ecological and public health issue.

2.2 Magnetic nanoparticle synthesis approaches

History has witnessed a comprehensive study of the advancement of various methods for MNP synthesis with the aim of producing MNPs of the required size, stability, biocompatibility, and morphology [14]. In the past, the role of MNPs has exponentially grown in different areas of physics, chemistry, contaminant sensing and detection, biology, medicine, etc [16]. Various noteworthy efforts have been devoted to the advancement of new approaches for MNP synthesis. The surface functionalization (FUCTN) of MNPs and the precision of their preparation are important, as they influence the physicochemical features, stability, movement, and contaminant sequestration efficiency of MNPs [16].

The two classes of NP synthesis approaches are the top-down (TD) and bottom-up (BOP) synthesis approaches. In a broad and general sense, MNP synthesis is grouped into three main procedures, namely the chemical, biological or microbial, and physical approaches [16, 17]. The TD synthesis approach (destructive approach) involves the breakdown of bulky molecules (bulk material) into smaller molecules and their subsequent transformation into NPs. Examples of the TD NP synthesis approach are the physical synthesis methods (mechanical crushing or milling, lithographic, physical vapor deposition, laser ablation, sputtering, and other destructive methods). The BOP approach (constructive approach) involves the formation of NPs from comparatively modest substances. This approach comprises the chemical and biological synthesis methods (chemical vapor deposition (CVD), the sol–gel method, spinning, spray pyrolysis, laser pyrolysis, coprecipitation, thermal decomposition, hydrothermal/solvothermal methods, non-thermal plasma techniques, and the use of microemulsions, polyol, microbes, and plant extracts), in which molecules/atoms are accreted to create various dimensions of NPs [16–18].

This approach is also used to synthesize NPs to overcome the productivity issues related to chemical and physical approaches owing to the duration and variety of steps involved in the synthesis process [15, 17, 19]. The green or biological synthesis approach is used in the synthesis of NPs, since it is environmentally friendly, a frugal and practicable method for large-scale production, and a cost-efficient approach that avoids the use of destructive and scarce chemicals. Such approaches are alternatives to the chemical route based on the application of plant extracts that have certain structures in their physical and chemical makeups [20]. This approach hinges on the bioreduction of metal salts into NPs, leveraging the numerous biomolecules (amino acids, phenolic acids, vitamins, proteins, and alkaloids) found in microorganisms (like fungi and bacteria) and plants (leaves, stems, roots, and fruit) or the interaction between compounds in the bioreaction mixture. Phenolic acids are common antioxidants which possess carboxyl and hydroxyl groups that can bind metals [20, 21].

The key biomolecules in NP synthesis using bacteria are proteins and polysaccharides. NPs can be synthesized using bacteria both intercellularly and extracellularly, with the extracellular synthesis approach being the most commonly used due to its straightforwardness. A key drawback of this approach is the separation of the biomass from the produced NPs and the extra precautions required when the biomass utilized may be pathogenic. Biomass seepage also poses an ecological problem, especially for biomasses that can be a source of disease. The NPs created via this process are often capped, making them inappropriate for sensing, electronic, and catalytic uses, which involve unrestricted surface ligands. In the fungal synthesis of NPs, fungi act as reducing agents (intracellular or extracellular) that produce extracellular proteins employed to stabilize NPs and induce biocompatibility. The key benefits of fungi over other microorganisms are their HM tolerance, easy management, and the production of extraordinary biomass [22].

Against this backdrop, the green synthesis approach represents an interesting option due to the peripheral carbon-based coating that takes place during MNP formation, which functions as a capping agent or reducing agent (terpenoids, flavonoids, tannins, and alkaloids) or as a stabilizing agent (carbonyl, amine, and carboxyl groups) during the synthesis of NPs. The green synthesis approach can increase MNP dispersity and chemical stability, improve their biocompatibility, and decrease the toxic impacts associated with other NPs coated with artificial carbon-based layers. Unexpected effects of the carbon-based layer formation in this synthesis method are the avoidance of unwanted NP aggregation and an enhancement of colloidal stability, which improve their effectiveness in aqueous systems [20, 22].

The physical approach comprises the TD and BOP methods. In the TD method, bulk materials are broken down into nanosized particles through high-energy ball grinding. Using this approach, NPs of the desired shape and size are not always obtained. In the physical BOP approaches (laser evaporation, wire explosion, and inert gas condensation), well-dispersed and correctly scaled tiny materials are achieved [14]. The physical approaches offer particular control over the size of NPs but may involve specific equipment and controlled settings. This comprises chemical reactions in a water-soluble (WS) monophasic liquid medium in which visible Fe hydroxide (OH) nucleus growth and nucleation are controlled. Also, the annealing temperature plays a critical part in MNP magnetization, as a temperature range of $9 \times 10^2 - 1 \times 10^3$ °C leads to better outcomes [15, 19].

The chemical synthesis approach (wet chemistry) is the most commonly utilized method due to the extraordinary yields that can be obtained. The chemical synthesis methods comprise chemical vapor deposition, molecular condensation, chemical reduction, sol–gel coprecipitation, thermal decomposition, and microemulsion and hydrothermal approaches. These approaches regularly take extended times, require the utilization of hazardous chemicals, and limit the amount of metals incorporated during MNP formation [22]. These approaches are based on NP precipitation from solution, and monodisperse particles are achievable if the precipitation profile complies with the LaMer and Dinegar model of standardized precipitation. According to this model, at a certain phase of supersaturation during precipitation from a solution, nucleation sites slowly grow in size through the redistribution of solutes in the solution toward the nuclei, and this occurs until monodisperse particles with specific sizes are achieved. The mechanism of this model [23–25] is illustrated in figure 2.1.

The most commonly utilized approach for MNP synthesis is the chemical synthesis approach, due to its extraordinary yields and maturity. The biological methods rank as the second most employed approaches. The physical approaches are the least utilized, even though they are gaining consideration in the research literature due to some of their advantages over traditional approaches [22]. MNPs have attracted more attention as efficient sorbents for the removal of various contaminants owing to their ease of separation, low cost, extraordinary SA, surface modifiability, outstanding magnetic properties, and high biocompatibility. The main routes employed for MNP synthesis are the coprecipitation, sol–gel, hydrothermal reaction, and microemulsion approaches [15, 19]. In the coprecipitation approach, MNPs are created from WS salt solutions in a passive environment with the addition of a base; this can take place at high temperatures or at room temperature [26].

Compared to the physical approaches, the chemical approaches are found to be more appropriate for the synthesis of MNPs, as they have broader applicability.

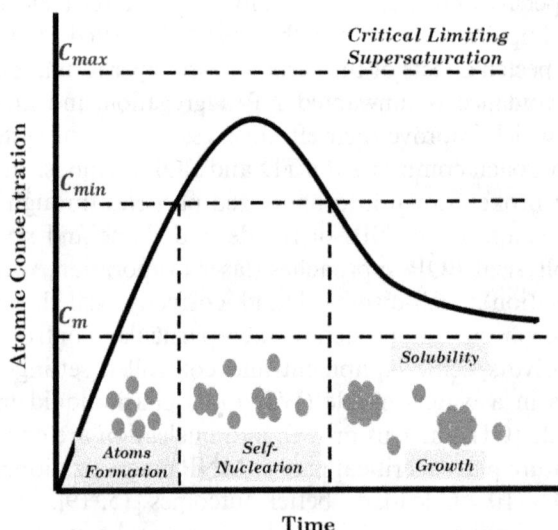

Figure 2.1. LaMer and Dinegar model of standardized precipitation.

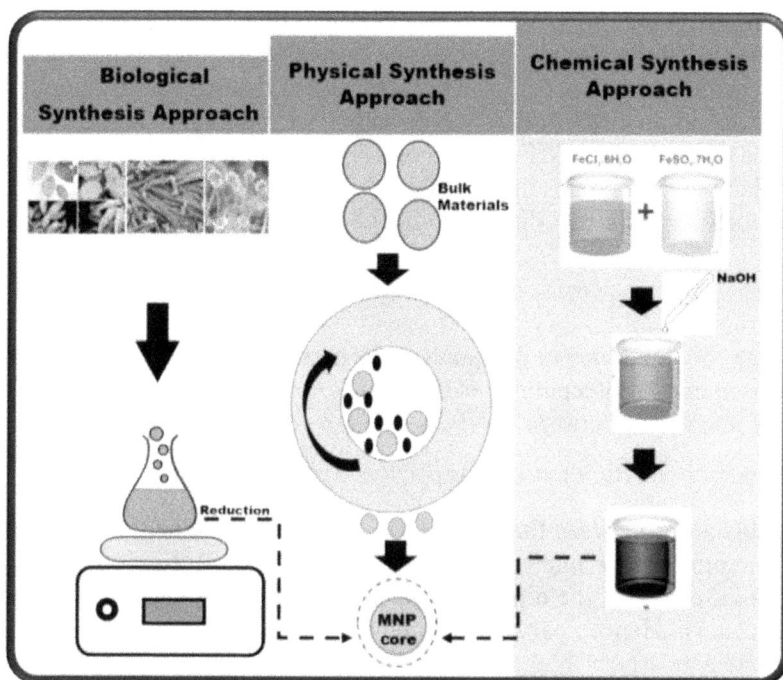

Figure 2.2. Various methods for MNP synthesis.

However, the various antagonistic effects of the chemical approach have been detailed, as noxious substances may be adsorbed and desorbed from NP surfaces created via the chemical method. Further key drawbacks of the chemical and physical approaches are their high costs and their utilization of toxic and harmful chemicals, which pose indisputable ecological threats and may be toxic to humans. Furthermore, owing to their level of scientific advancement, the chemical and physical approaches are not widely utilized these days. Research efforts are more focused on biological approaches [16]. Additionally, rare materials utilized in the biological synthesis approaches (concentrates of plants and microorganisms) are readily accessible compared to the rare materials used in the chemical synthesis approaches [16]. The various synthesis approaches are shown in figure 2.2.

2.3 Functionalization of magnetic nanoparticles

Functionalization is the process of altering the surfaces of NPs with various molecules to improve their properties for different uses. In this process, organic or INORG molecules are attached to the NP's surface, thereby improving stability, preventing aggregation, and enabling precise interactions [27]. Functionalizing MNPs enhances their sorption properties and catalytic efficacy. They can readily be removed from a medium by an EMF [27]. Superparamagnetic materials are magnetized when exposed to an externally applied MF but display no hysteresis in their magnetization curve and have zero magnetization at a practical field of zero. In most situations, FeO NPs have been broadly utilized as choice materials for environmental as well as other applications. Furthermore, their application requires NPs to be chemically stable, consistent in size, and well-dispersed in a liquid medium [28]. However, in an EMF, these MNPs acquire a dipole moment that is proportional to the strength of the MF. The dipole–dipole interaction induces the clustering of NPs, which creates chains and associated structures. In the presence of an MF, these NP domains acquire a dipole moment according to equation (2.1):

$$\mu = \frac{4}{3}\pi r^3 \mu_o \chi H \tag{2.1}$$

where r, μ_o, χ, and H represent the radius of the particle, magnetic permeability (μ) in a vacuum, the magnetic susceptibility of the particle, and the EMF strength, respectively. The energy interaction between MNPs is found by employing equation (2.2):

$$U(r, \theta) = (\mu^2/4\pi\mu_o)[1 - 3]\cos^2(\theta)]/r^3 \tag{2.2}$$

where θ is the angle between the applied MF and the axis. When the thermal energy (kT) is overcome by the interaction energy of the induced dipoles, aggregation of particles takes place via the dipole–dipole interactions, leading to the formation of magnetic chains. Hence, particles should be stabilized or encapsulated with a surfactant at a relatively high concentration to avoid buildup. However, the use of huge quantities of surfactants may significantly hinder the medical, biological, and environmental applications of MNPs [28].

One of the approaches for MNP surface FUCTN is the one-pot *in situ* surface FUCTN. Employing this approach alongside FUCTN allows a narrower particle size distribution to be achieved. This involves the simultaneous feeding of the MNP precursor alongside the coating material into the reaction mixture. The coating process is initiated as nucleation ensues, thereby further stopping the growth of particles. Another approach extensively used for the FUCTN of MNPs is the post-synthesis surface coating approach. In this method, the complete process involves the synthesis of MNPs and the modification of their surfaces. Bifunctional compounds are utilized for this, in which a chelating binding group is first reacted, and successive coupling site group modification leads to the final functional group. In this approach, the created MNP surface is covered with various functional materials (FMs) attached by chemical or physical links such as covalent bonds, hydrogen bonds, affinity interactions, electrostatic forces, etc [29].

The mechanism of MNP surface modification usually comprises the addition of a ligand, ligand exchange (LEx), and encapsulation. The surface modification of MNPs through the addition of ligands involves the addition of ligands to MNPs' external surfaces without removing the previous ones. These are added directly and are adsorbed on the surface of the MNPs, interacting through a hydrophobic (HP) or covalent process with the MNPs' surfaces [27]. The four routes of ligand addition include ligand addition on the MNPs' surface without a capping agent, INORG layer growth on the surface of the MNPs with ionic or other non-explicit interactions (indirect ligand addition), covalent bond formation between the new ligand and the current ligand, and the manipulation of HP species into the MNPs' hydrocarbon shells capped by ligands. The use of LEx on MNPs involves the replacement of specific HP ligands with added hydrophilic ligands (HLs) that have stronger binding properties. MNPs move from the organic phase to the aqueous solution as a result of LEx. In the encapsulation of MNPs, HP ligands on MNPs are coated with an amphiphilic substance. This is accompanied by positioning the HL constituent of the materials in the solution and adding the HP component of the materials with the primary ligands on the surface of the MNPs. The mechanism of the surface FUCTN of MNPs is depicted in figures 2.3(a)–(c) [29].

Different materials can be employed to meet the mandatory conditions for the use of MNPs. These materials include carbon-based materials (polymers, small molecules, surfactants, and monomers) and INORG materials (metal oxides, silica, non-metals, sulfides, and metals). Organic materials are often used to passivate NP surfaces during or after the preparation process to avoid aggregation. In the absence of any proper surface covering, NPs have HP surfaces with an extraordinary SA-to-volume ratio. NPs aggregate and form bulky clusters, which result in enhanced particle sizes due to the HP interaction between NPs. In the FUCTN of MNPs with organic materials, functional groups (FGs) such as amino, hydroxyl, aldehyde, and carboxyl groups are affixed to NPs. These groups can connect with energetic biosubstances such as enzymes and DNA during the treatment of WWT containing HMs and dyes. Coating these NPs with organic materials does not negate their properties and simultaneously preserves the NPs' magnetic properties. Figure 2.4 shows three types of nanostructures functionalized (FZD) by organic materials: (1) the core–shell structure: a core of FeO particles covered by a layer of biological compounds (the shell); (2) the matrix structure, consisting of (a) a mosaic: organic molecules forming a shell coating glazed on homogeneous FeO NPs, or (b) a shell–core: a shell of FeO with a core of organic material; and (3) the shell$_a$–core–shell$_b$ structure: a shell coating made of organic compounds covering a structurally FZD FeO NP, which itself consists of an FZD structure of Fe_3O_4 NPs [27, 30].

The surface features of NPs FZD with trivial molecules or surfactants are grouped into three kinds, namely: (1) oil-soluble (FZD NP surface to increase NP stability and thus yield monodispersity; comprises molecules with weak attraction for solvent settings; generally includes HP groups of fatty acids, alkyl phenol), (2) WS (FZD NP surface applied in biodetection and bioseparation; comprises chemical groups with a strong attraction for solvent settings; includes HP groups of ammonium salts, polyol, and lysine), and (3) amphiphilic (FZD NP surface comprises HP and HL chemical groups shown by the main chain of these FZD NPs, which have both water- and oil-solubility; an example is sulfuric lysine) [30].

The attachment of INORG small compounds on the surface of FeO NPs is the informal and first practical modification method. INORG materials utilized to modify the surface of FeO NPs include silica, non-metals, metals, metal oxides, graphene, and sulfides. These greatly improve the properties of bare FeO NPs [31, 32]. The two key surface modifications of FeO NPs generally adopted are the synchronized addition of two or more INORG trivial molecular compounds into a solution comprising the FeO NPs. The INORG small compounds interact with one

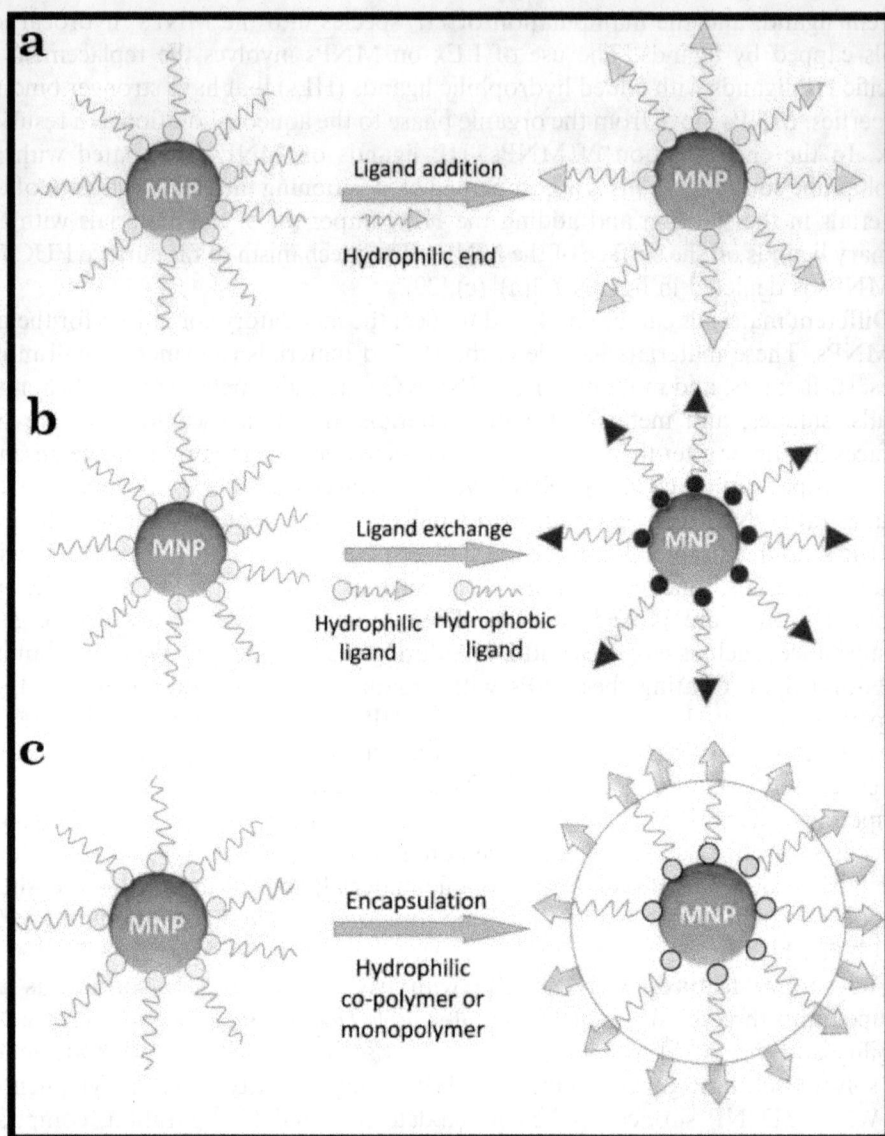

Figure 2.3. Mechanisms for the surface FUCTN of MNPs through the (a) addition of a ligand, (b) LEx, and (c) encapsulation. Reprinted from [29], Copyright (2023), with permission from Elsevier.

another and precipitate on the FeO NPs' surface. In this approach, it is difficult to quantitatively control the INORG material deposition thickness on the surface of the NPs. Furthermore, a critical issue in this approach is the weak bonding strength between NPs and the INORG compound, which limits their use in WWR treatment. The second surface modification of FeO NPs using INORG compounds is the deposition of the INORG metal compound on the NPs' surface using the vapor phase coprecipitation approach or the microemulsion approach. Magnetic

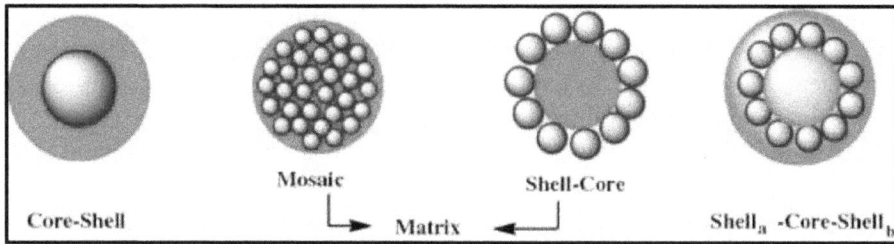

Figure 2.4. The core–shell, matrix, and shell$_a$–core–shell$_b$ of the FZD structure of NPs by the organic materials. Reproduced from [30], CC BY 2.0.

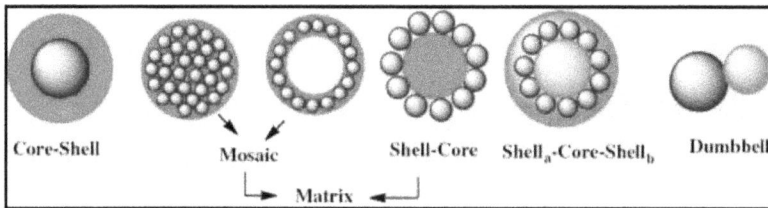

Figure 2.5. The structures of FeO NPs FUCTN with INORG compounds. Reproduced from [30], CC BY 2.0.

composite NPs with an embedded or core–shell structure are synthesized using this approach. The covering thickness of the INORG compounds on the MNPs can easily be controlled, and a microporous or mesoporous structure is created on the NP surface. It has been reported that a high bonding strength results from the chemical bonding interaction on the boundary between the INORG compounds and the NPs. Hence, in WWR remediation, the modified Fe_3O_4 NPs are stable and have improved adsorption (ADPN) performance for contaminant removal. Owing to the controllability, stability, and good ADPN performance of FeO NPs when modified using direct deposition, this technology is generally used in the FUCTN of MNPs and the treatment of WWR. The approach of attaching INORG small compounds to NP surfaces is modest and appropriate in operation. After the surface modification, the structural stability of the MNPs is improved. Yet, the use of INORG small compounds to coat MNPs for contaminant removal is only a physical ADPN method, and these adsorbents lack abundant specific FGs which can interact with organic compounds, thereby leading to low effectiveness in the elimination of WWR contaminants. Based on the above constraint, in real industrial WWT procedures, organic compounds are employed to further modify the INORG material surface to ensure better removal performance for contaminants [31].

Figure 2.5 shows the different matrix-dispersed FeO NPs produced via the FUCTN of NPs using INORG compounds. The combination of hollow silica spheres with FeO NPs generally leads to a mosaic structure, while the shell–core structure is created by individual FeO NPs associated with their internal coating, according to Wu *et al* [30]. However, the shell$_a$–core–shell$_b$ form of composite NP is obtained using coating technology. Shell$_a$ may be made from metal NPs, quantum dots, or a polymer. Similarly, shell$_b$ can be made from identical or different FMs.

This kind of composite NP is anticipated to significantly increase the range of FeO NP applications. Dumbbell structures are generally created via the epitaxial development of FeO NPs or INORG NPs on INORG material seeds or FeO NPs; such structures allow bifunctional composite NPs to be produced.

Coatings made from organic or INORG compounds provide stability and help to prevent aggregation. Occasionally, different FGs utilized for surface activation may interfere with the particles' magnetic properties, in turn impacting their effectiveness. MNP surfaces are covered to improve their usefulness in targeting and attaching to selected contaminants with elevated selectivity and specificity. The small volumes of materials used in MNP surface modification avoid the creation of secondary pollutants during the remediation procedure and regeneration, thus mitigating their shortcomings [33].

2.4 MNPs—important properties and characterization

Their physical and chemical properties of MNPs strongly influence their precision and effectiveness in WWT. Their overall excellent performance is based on their chemical, electrical, magnetic, physical, and structural characteristics, which improve their breadth of application. Their regeneration, reusability, easy separation, and lower operational expenses are some of the advantages of MNPs. The characterization of MNPs is generally associated with challenges such as the lack of reference materials for equipment calibration, the interpretation of results (complicated by the multidisciplinary attributes of this field), and complications linked with sample preparation. There is no existing standard procedure for the characterization of NPs. Effective and robust measurement approaches for NPs have a key effect on the uptake of these materials in systematic uses [16, 34].

Particles characteristically ranging in size from 1 nm to several μm play a critical role in the development of nanoscience and NT. Nanosized particles have physicochemical properties that are neither features of the atomic scale nor their bulk equivalents. The magnetic properties, superparamagnetic phenomena, and quantum tunneling of magnetization exhibited by MNPs are a result of the quantum size effect and the huge SA. This is attributed to individual particles acting as a single magnetic domain [35].

Accurate MNP characterization is crucial to achieving reproducible results. The composition, shape, and size also determine MNPs' magnetic characteristics. Characteristically, the morphology and size, surface chemistry and charge, crystal structure, chemical composition, and magnetic behavior are some of the properties of NPs that are considered for analytical use. The effectiveness and magnetic properties of MNPs in various applications are immensely reliant on their surface functional groups, structure, morphology, and size. For this reason, various characterization approaches are used [16, 34]. For instance, the reduction of MNPs' size leads to exceptional magnetism, while the MNPs' shape affects inhomogeneity. The validity and reliability of different characterization approaches are also critical in the assessment of the most effective synthesis approach, which may improve the effectiveness of MNPs. Some important features of MNPs were

Table 2.1. Types and properties of FeO.

FeO chemical formula	Color	Crystal system	Form of magnetism
γ-Fe$_2$O$_3$	Red-brown	Tetragonal or cubic	Ferrimagnetic
Fe$_3$O$_4$	Black	Cubic	Ferrimagnetic
α-Fe$_2$O$_3$	Red	Rhombohedral hexagonal	Weakly ferromagnetic

discussed in the preceding subsections [16, 34]. MNPs' physicochemical and magnetic properties [27] are shown in table 2.1.

2.4.1 Crystallographic analysis of MNPs using x-ray diffraction

The magnetic properties of MNPs are strongly affected by the degree of the component mineral's crystallinity and purity. The structure of NPs can be determined via infrared (IR), thermal analysis, and x-ray diffraction (XRD) spectroscopy. The magnetic property of MNPs has a critical effect on their effectiveness, separation, stability, recovery, and recyclability. Ferromagnetic particles are MNPs which show permanent magnetization after being detached from an MF, whereas no permanent magnetization is observed in paramagnetic (PMC) particles when removed from an MF. NPs' magnetism or magnetic properties hinge on the magnetic induction (B) and MF strength (H). Overall, there is a linear association between B and H, which is governed by the NPs' μ value. MNPs exhibit PMC properties if $\mu > 1$ and diamagnetic (DMC) properties if $\mu < 1$. The magnetic properties also limit the requirement to separate the catalyst by centrifugation, filtering, or extraction processes. Furthermore, the MNPs' surface can meaningfully decrease dispersion and hence product purity [16].

The study of crystal structures in materials analysis is known as crystallography. It is explored for both quantitative and qualitative analysis of materials [17]. Various structures in the same material can have a substantial impact on key features. Due to their extraordinary proportion of surface atoms, NPs may have novel phases, which may affect their size. Also, NPs may lack any regular structure (amorphous), they may be crystalline, or they may have non-crystallographic quasi-close packing. They can also be a sole crystallite or composed of smaller crystal grains. Characteristic approaches for assessing the materials' crystal structures comprise XRD and transmission electron microscopy (TEM) [22].

Powered XRD is important in assessing NMs' crystalline nature and structure. It offers an understanding of the diameter, crystallinity, and structural changes caused by coatings, immobilized materials, and the FUCTN of MNPs. In the XRD analysis of materials, a detailed structural scattering of x-rays (XRs) takes place in certain directions when a crystalline material diffracts an incident XR. Assessment of the precise angles where the XRs are sensed permits crystal plane separation and calculation using Bragg's equation. The crystal phases present in the material, the extent of alignment, and the crystal grain size can be identified/assessed via this process [36]. However, XRD alone may not differentiate Fe$_3$O$_4$ NPs from γ-Fe$_2$O$_3$

owing to the similar patterns created by identical cubic spinel structures [36]. XRD analysis was used to carry out a structural analysis of the effectiveness of γ-Fe_2O_3 NPs fused on activated carbon (AC) in the removal of dye molecules (figure 2.6). Based on the study result, strong XRD peaks caused by the nanomaterial were observed at 2θ values of 30.1°, 35.7°, 53.1°, 57.1°, and 62.7°. These were ascribed to the lattice planes of 220, 311, 422, 511, and 440 in the cubic structure of the γ-Fe_2O_3 NPs (JCPDS, No. 04-0755). Based on the Debye–Scherrer formula (DSF), the material's crystal size was assessed to be 13 nm based on the full width at half maximum (FWHM) of the 311 peaks [37].

2.4.2 Transmission electron microscopy

MNP shape and size are physicochemical properties that help to ensure MNP stability. The particle size of MNPs is greatly impacted by the magnetic moment and response. As reported, the saturation magnetization of Fe_3O_4 particles declines as the NP size decreases. Similarly, a reduced NP size leads to an increase in the surface area, which may impact the MNP's non-crystalline nature, and thus their magnetic moments may also be affected. Tiny MNPs can exhibit superparamagnetic properties, with improved magnetization in contrast to PMC materials. High-resolution TEM (HRTEM) and field emission scanning electron microscopy (FESEM) are utilized to characterize NP sizes and shapes. NPs' atomic arrangement and interfacial structures can also be evaluated by employing these approaches [16].

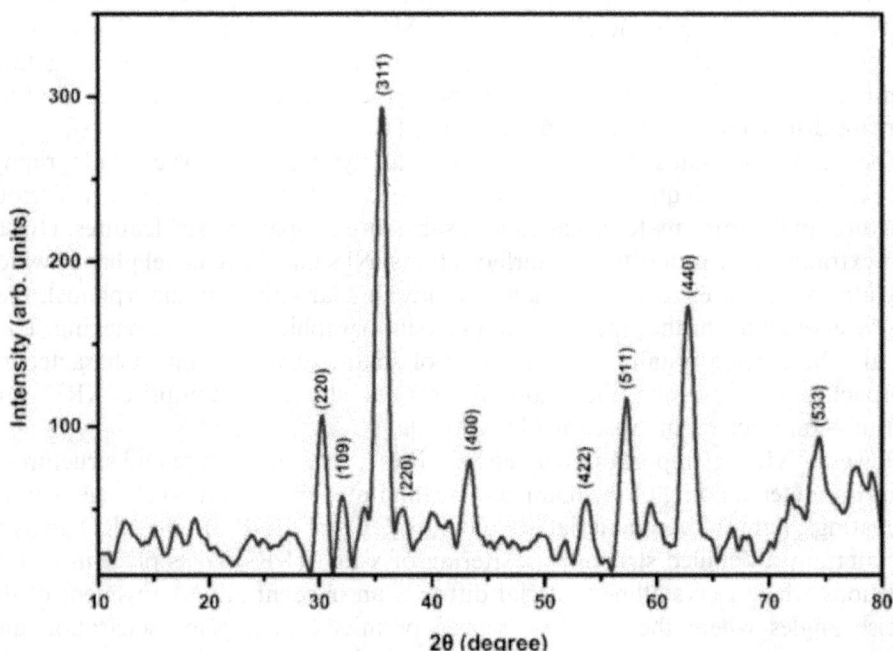

Figure 2.6. XRD pattern of AC/γ-Fe_2O_3 nanocomposites (NCs). Reproduced with permission from [37]. © 2016 Elsevier B.V.

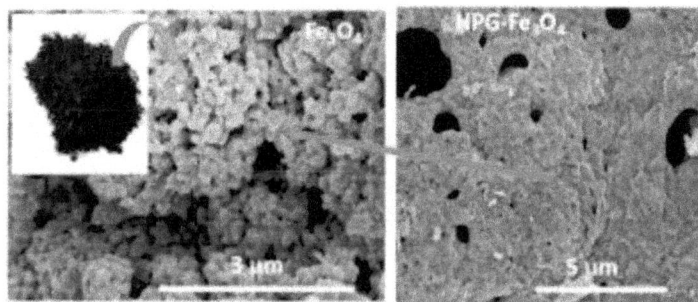

Figure 2.7. HRTEM image of magnetic NCs. Reprinted from [40], Copyright (2020), with permission from Elsevier.

Electron microscopy offers high-resolution images and chemical analysis, which enable scientists to detect nanomaterials (NMs) at the atomic scale [38]. It is used to research material particle size at the NS and the crystal structure of particles [17]. It is the preferred approach for analyzing the shape, size, and homogeneity of produced MNPs. However, it is restricted by orientation effects that lead to ambiguous images and issues associated with a large number of particle quantifications. Sample preparation protocols are crucial in obtaining dependable results from TEM images, and these should not lead to changes in samples. High-resolution TEM (HRTEM) is a phase contrast imaging mode of TEM in which images are created by merging both scattered and transmitted electrons. A larger objective aperture is required for collecting the scattered electrons in HRTEM compared to conventional TEM, which provides the numerical calculation of the size and morphology of NMs. HRTEM also provides appropriate resolution for the imaging of single-particle crystal structures and hence provides critical information about NMs' structures [34]. Good resolution that does not need correction is obtained by analyzing sufficiently thin samples (<200 nm) using TEM [39]. In this study, a prepared magnetic nanocomposite morphology used for HM removal was assessed using HRTEM. The NCs observed in this study were found to have a spherical morphology, with a core–shell structure type. The particle size of the NC core was found to be 21–30 nm, as shown in figure 2.7 [40].

2.4.3 Scanning electron microscopy (size and morphology)

Scanning electron microscopy (SEM) is used to obtain more detailed information about the nature of NPs. It offers a higher resolution than that of optical microscopy. It was introduced to study the particle morphologies of different samples. This microscopy is extensively used to assess the size distribution, shape, and morphology of the produced micro- to NS materials. When used in combination with energy-dispersive XR spectroscopy (EDX), sizes, shapes, and elemental compositions or chemical identities can be obtained [41]. This approach reports the total size of the particle; hence it is suitable for NP core–shell characterization. A key constraint of this approach is that it is not appropriate for the detection of NMs with sizes below 20 nm. The images produced are created utilizing SEM, in which electrons are employed as an alternative to light [17, 34]. Low-resolution SEM has some advantages, such as improved XR spatial resolution and a correction procedure for impenetrable samples [39].

2.4.4 Energy-dispersive x-rays

EDX and energy-dispersive spectroscopy (EDS) are employed to detect the chemical compositions of prepared MNPs. These approaches are essential for confirming the material's successful modification and assessing the effects of surface modification on MNPs [36]. EDX microanalysis is an approach to the study of elements. It is linked to electron microscopy and is based on the generation of distinctive XRs that show the presence of elements in samples. Two important physical events, elastic scattering (ES) and inelastic scattering (IS), occur when XRs are incident upon atoms. ES results in a change in electron direction and a measurable loss of energy, mostly triggered by the nucleus's interaction with the XRs. IS involves an interaction with bound electrons and atomic nuclei, resulting in energy loss but no detectable change in the electron direction [42].

The key factor determining the shape of the interaction volume is ES, while the length of the interaction time depends on IS. However, atoms are ionized when they return to their ground state, resulting in the emission of characteristic XRs. The XR photon energy is the potential energy that results from the difference between the two orbitals involved in the transition and thus depends on the element. A photon-energy-sensitive detector measures the XR emission at various wavelengths. The XRs are characteristic of the elements from which they were emitted, and they provide data on the elements present in the sample. Certainly, their energy depends on the atomic number (AN) of the element involved in the interaction. The EDX detector system provides an instantaneous display of all low-energy (1–20 keV) XRs collected during any specific study period, and the XR energy is represented as a spectrum, which is a histogram plot of the number of counts versus x-ray energy. The spectrum comprises semiquantitative and -qualitative information. The peak position in the spectrum and its energy identify the element, while the peak area is proportional to the number of atoms of the element in the x-rayed area. Based on the qualitative assessment, elements are identified in the spectrum using the manufacturer's software. Given that each element has a unique energy signature, the fingerprint or signature of the spectrum can be related to the reference spectrum. This analysis is employed for elements with ANs above 10. A key limitation of this approach is its inability to distinguish between nonionic and ionic species. It is applied in environmental contamination studies and offers significant benefits in detecting HM contamination [39, 42].

2.4.5 Infrared spectroscopy

IR spectroscopy is utilized to assess FGs and their probable interactions, thus providing an understanding of MNP development and their surface FUCTN [36]. A wide-ranging spectrum of molecular systematic tools based on IR spectroscopy was introduced in the 1950s, and in the 1970s, further improvements to IR tools led scientists to advance IR spectroscopy based on a novel computational study, which led Helm and Naumann to introduce this approach for the *in situ* assessment of bacteria [43].

IR fundamentals are based on emitted radiation from an IR source, which passes through an interferometer made up of a beam splitter, a moving mirror, and a static mirror. The interferometer assesses the emitted light wavelength using the pattern of interference, which improves precision. The IR spectra are acquired by exposing a sample to IR radiation and then assessing the transmitted radiation intensity at a certain wave number. The scan number can be adjusted based on the required sample analysis quality, and the most common scan number utilized is 2^8. Certain molecular groups can be distinguished at certain wave numbers using IR irradiation. The x-axis and y-axis of the spectrum denote the wave number and the transmittance or absorbance, respectively [43].

Overall, the intensity of the absorption bands detected in the attenuated total reflectance (ATR) IR spectrum is less than the intensity of the transmitted IR spectrum. IR microspectroscopy is an IR approach that combines a microscope and IR spectrometer to obtain spatial and chemical spectral information instantaneously. The key difference between IR and ATR-IR is the penetration depth. Generally, the bulk properties of samples are measured by transmittance IR, and in this case, the IR beam can only probe samples with a thickness of about 300 nm. The two approaches also differ in their requirements for sample preparation [43].

The key benefits of using this type of spectroscopy are its time-efficient approach owing to the simple preparation of samples, the short time needed to complete the spectral analysis, and the fact that samples can be assessed in various forms (solids or liquids). Additionally, only small sample volumes are needed for analysis, and these generally fall within the microliter (μl) scale for liquids or microgram (μg) scale for solids. This approach is generally not damaging. Other outstanding benefits of IR are its elevated signal-to-noise ratio (possibly owing to the instantaneous assessment of wavelength – see Fellgett's advantage), elevated energy throughput (due to the prevention of light dispersion in IR, which causes elevated energy throughput – see the Jacquinot advantage), and huge stability and precision (the outstanding advantage of the IR method is the utilization of a helium–neon (He–Ne) laser, which functions as an internal reference for individual scans and offers stable and precise wave number scales for an interferometer – see the Connes advantage). Some of the disadvantages of this approach include various background scans and sample scans needed to circumvent artifacts and differences in the spectra owing to adjacent environmental influences on sample heterogeneity. To purify the sample, sample pretreatment may be needed to avoid peaks from overlapping in the spectra, and the raw data can require wide-ranging postprocessing [43].

Based on an IR analysis of prepared NCs utilized for dye molecule sorption (figure 2.8), the extensive absorption peaks noted at 529 and 345 cm^{-1} were attributed to Fe-O stretching and the bending vibration modes of γ-Fe$_2$O$_3$ NPs. Bands observed in the 1500–3500 cm^{-1} range were attributed to the sorbed water molecules in the KBr medium. The band noted at 1100 cm^{-1} was attributed to the

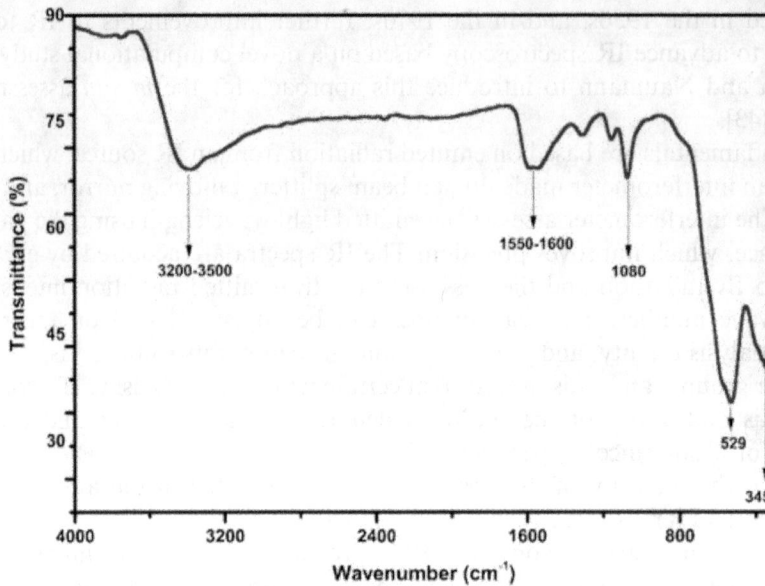

Figure 2.8. IR spectra of γ-Fe$_2$O$_3$ NPs fused on activated carbon (AC). Reproduced with permission from [37], © 2016 Elsevier B.V.

vibrational mode of sorbed CO$_2$ on the NPs' surface [37]. Based on a study by Liu *et al*, IR analysis was carried out on synthesized magnetic NCs (figure 2.9). According to the study results, a wide peak detected at 586 cm^{-1} was due to the vibration absorption peak of FeO. Peaks at 1043, 1318, 1278, 1668, and 2939 cm^{-1} were attributed to the stretching vibration peaks of Si–O–Si and C–N in amide, amide III bands creating coupling between δ_{N-H} and v_{C-H} in amide, the stretching vibration peak of amide C=O (amide I band), and the symmetric or antisymmetric C–H stretching vibration of methylene [44].

2.4.6 UV–vis spectrophotometry

A fundamental recognition approach in many frontier research fields such as materials synthesis, environmental contamination detection, biological detection, and mineral collection is solid-state UV–vis ADPN spectroscopy. It is extensively applied in semiconductor studies for the detection of band-edge structure, the E$_g$ value, and novel band creation within the forbidden gap. This approach has been further blended with synchrotron XRD spectroscopy, Raman scattering spectroscopy, photoluminescence spectroscopy, etc [45]. This approach is extensively applied for NP characterization. It is used for quantitative investigations of NP concentration based on the Beer–Lambert law, which associates absorbance with concentration (equation (2.3)). This approach depends on the interaction between the NPs and light, which leads to variations in the absorption spectra. Light absorption by NPs leads to the presence of distinctive peaks in the UV–vis spectrum [15]:

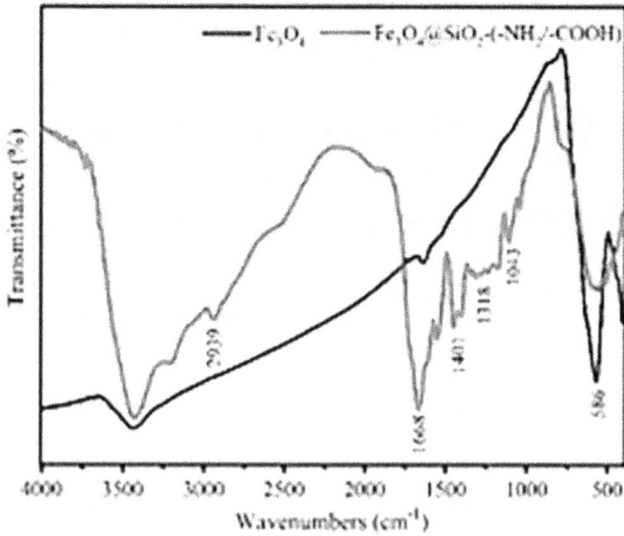

Figure 2.9. IR spectra of magnetic NCs employed for the removal of HMs. Reproduced with permission from [44]. © Copyright © 2012 Published by Elsevier Ltd.

$$I = I_o \cdot e^{-\alpha d} \tag{2.3}$$

where I, I_o, α, and d are the intensity of the transmitted/reflected light, the intensity of the incident light, the absorption coefficient, and the thickness of the sample (applicable to thin films), respectively [46]. It is expected that the sample being studied should be standardized and placed on a clear substrate. In this situation, data about the sample absorbance spectrum is provided by the degree of transmittance. Reflectance must be used when the sample is placed on an impervious (non-transparent) substrate [46].

Optical spectra reflect the intensity of electromagnetic radiation. They are presented in wavelength or frequency order and reflect distinct energy bands. Figure 2.10(a) shows the wavelength order; electromagnetic waves are classified as γ-rays, XRs, UV–vis light, IR light, microwaves, and radio waves. They are produced by the transitions of particles between various energy stages and reflect the distinct interaction between electromagnetic waves and matter. Under light excitation, carriers in the ground state absorb light quanta that cause their energy state to be raised to various excited states. The residual light quanta pass through the material without interaction, and thus the detected transmitted light can provide information about the absorption of the incident light ray at various wavelengths; hence, it is known as absorption spectroscopy. The profile of the absorption spectroscopy is dependent on the material's intrinsic band structure. Organic molecule absorption in the UV–vis band generally involves π (doublebond), σ (singlebond), and n (nobond) interactions (figure 2.10 (b)). The double- and single-bond electron orbitals are categorized as antibonding orbitals (*). The four forms of the permitted electronic level orbitals are $\sigma \rightarrow \sigma^*$, $n \rightarrow$

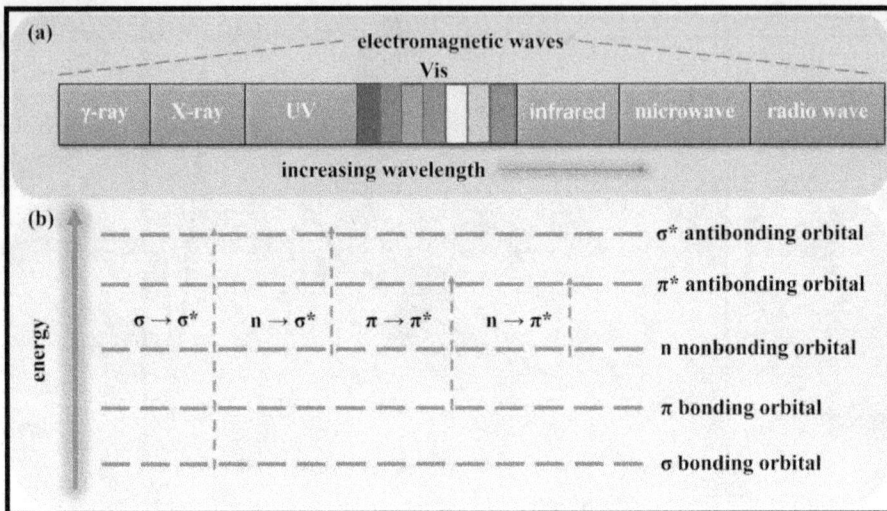

Figure 2.10. (a) Electromagnetic waves with increasing wavelengths and (b) the orbital transitions of the electronic energy level. Reproduced from [45]. CC BY 4.0.

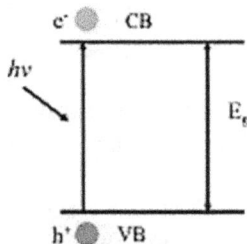

Figure 2.11. Electron optical transition from the valence and conductance bands of MNPs induced by photon absorption. Reprinted by permission from Springer Nature Customer Service Centre GmbH: [Springer] [Handbook of Nanofibers] [46], Copyright (2019)

σ^*, $\pi \rightarrow \pi^*$, and $n \rightarrow \pi^*$. Based on these transitions, the materials' photoelectric properties are assessed [45].

The bandgap (or E_g forbidden bandwidth) of MNPs is a vital optical factor, which can be determined using UV–vis spectrophotometry. It is the nominal energy of an absorbed photon required to create a free-electron–hole pair in the conduction and valence bands of MNPs (figure 2.11). Owing to structural defects and hanging surface bonds, electronic states can appear near the conduction and valence bands. They have a nonlinear distribution in E_g. Hence, an optical transition takes place between states, characterized by the energy E_g^*. The condition $E_g > E_g^*$ is known as the Urbach tail (figure 2.12). The Urbach tail can indicate the number of flaws and their distribution in E_g [46].

Semiconductors, insulators, and metal conductors are grouped using increasing values of E_g. The E_g value and the band-edge structure play fundamental roles in the excitation, movement, and successive changes of photogenerated electron processes.

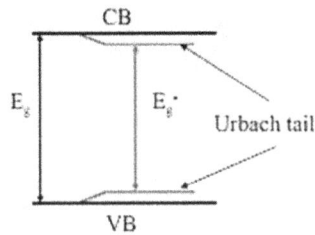

Figure 2.12. Urbach tail formation in the E_g of MNPs. Reprinted by permission from Springer Nature Customer Service Centre GmbH: [Springer] [Handbook of Nanofibers] [46], Copyright (2019)

Based on these impacts, a practicable approach to regulating and moderating the materials' macroscopic electrical, photoelectric, and optical properties is E_g engineering [45].

2.5 HMs

HMs are introduced into water resources by various manufacturing wastes such as the wastes from electronic devices and metal manufacturing, mining, power generation, and metal plating facilities [40]. Through various point sources (PSs) and non-point sources (NPSs), such as coal mining, domestic waste, agricultural activity, and the leather industry, HM ions are released into water resources. PSs are specific classes of contamination that cause elevated levels of HM ion pollution in water and arise from both industrial and domestic sources. NPSs produce contributions that emanate from extensive sources, as opposed to intensive sources at defined sites, and are considered to be of negligible consequence compared to the contributions from sources that are concentrated in an exact site. Examples of NPSs include the discharge and flow of detrimental pollutants from agricultural land, landfills, and dumping sites into water resources [47].

The primary sources of industrial WWR are paper and pulp producers, tanneries, chemical and petrochemical plants, food processing, and other manufacturing businesses. These WWRs are characterized by elevated organic concentrations $(1–200 \text{ g l}^{-1})$, non-neutral pH values, diverse temperatures, and high HM content, salinity, and turbidity. Similarly, WWR produced by petroleum refining, textile processing, leather production, and food preservation and processing may contain elevated salt concentrations. The composition of WWR varies, depending on the chemicals utilized in the upstream processes and the nature of the treatment it has undergone; hence, cataloging industrial WWR into detailed classes is thought-provoking [2]. Industries consume substantial amounts of water; on an international scale, they use approximately 22% of the total water in high-revenue nations, but this figure can reach 60%. By 2050, manufacturing industries alone are projected to increase their water consumption by 400% [2].

The persistent pollution caused by metal ions (MIs) entering external and underground water systems alters the fauna and flora of hydrological systems. Certain HMs are not biodegradable, and they tend to bioaccumulate, which is readily detectable in living organisms as an increase in their concentrations. They are also relentless and can directly or indirectly impact different organisms owing to biomagnification. They are extremely noxious and carcinogenic substances [48], as shown in figure 2.13 and table 2.2.

HMIs are of distinct consequence owing to their persistence in the ecosystem, long-term buildup, and potential danger to human health through the food chain [49]. When they enter the food chain, they can accumulate at low concentrations in existing organisms and are likely to cause antagonistic impacts such as nervous system impairment, kidney failure, and tumors. When they attach to living organisms' proteins, small metabolites, and nucleic acids, this results in the loss or alteration of biological functions or the perturbation of the organism's metal control sites. MIs are also responsible for genetic modification, which can persist over many generations. They can also impact marine organisms (fish, phytoplankton, and zooplankton) accumulating in some organs and triggering oxidative impairment, damage to the immune system (affecting the survival and growth of marine organisms), and endocrine disruption. HM contaminants such as cadmium (Cd), copper (Cu), arsenic (As), mercury (Hg), chromium (Cr), zinc (Zn), and nickel (Ni) have a comparatively elevated carcinogenic potential [4, 40, 48, 50].

The discharge of WWR from various manufacturing processes causes elevated concentrations of dissolved metals and dyes, which must be decreased according to legislative standards. To eliminate these threats, various approaches for water remediation have been utilized. An effective technology for decreasing the concentration of HMs and dyes is the use of ADPN based on various adsorbent types. Due to advancements in NT, NS materials have been developed for the treatment of WWR [4, 51]. The emphasis on the removal of dissolved contaminants from water is a key concern for society, since contaminants represent hazards to both public and environmental health. HMs are toxic and carcinogenic, and they can easily enter the food chain. According to the Environmental Protection Agency (EPA) of the United States, HMs are considered priority contaminants that must be removed from or

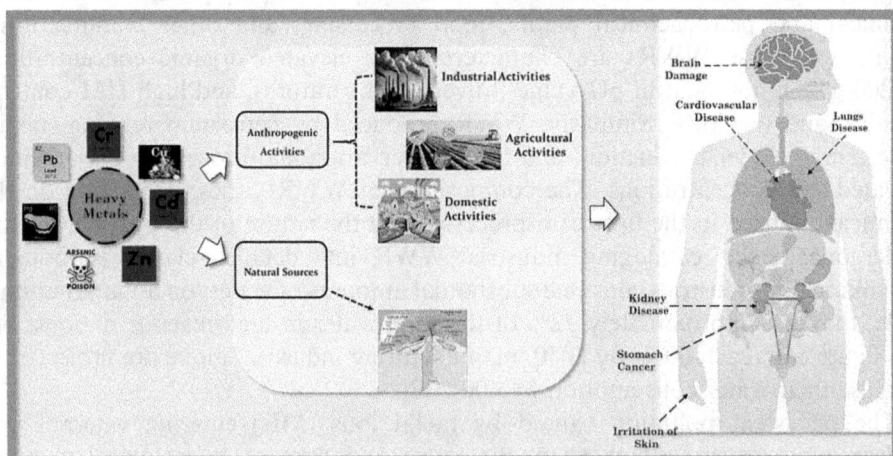

Figure 2.13. Sources of HMs in water and their effects on human organs.

Table 2.2. Origins of HM contamination in the environment, key organs and effects, and the permissible levels of HMs.

HM	Human-induced sources	Key organs and effects	Maximum permitted level (μg l^{-1})
Pb	Pb-based batteries, cable sheathing, pigments, alloys, glazes, plastic stabilizers, ammunition, rust inhibitors, solder, pesticides, coal, gasoline, fertilizers, emissions from vehicles, and metal purification	Cardiovascular and reproductive systems, lungs, bones, immunological and hematological systems, anemia, liver, spleen, gastrointestinal issues, brain, mental retardation in infants, nervous system disorders, and kidneys	10
As	Glass and electronics production, fly ash, irrigation with municipal effluents, industrial effluents, livestock manure composts, sewage sludge, and pesticides	Immunological and gastrointestinal systems, brain, kidneys, cardiovascular and metabolic systems, endocrine system, lungs, and skin.	10
Cu	Pesticides, fertilizers, corroded plumbing systems, electronics and cable industry, steel industries, chemical manufacturing, electroplating, mining industries, and metallurgy	Immunological and gastrointestinal systems, hematological systems, lungs, kidneys, cornea, and liver.	2000
Cr	Leather and textile industries, tanning and electroplating industries, anticorrosive products, steel and pipe mills, and industrial sewage.	Taste, gastrointestinal and reproductive systems, lung tumors, pancreas, skin, kidneys, liver, brain, and lungs	50
Cd	Metallurgical, plastic, and electroplating industries, insecticides, synthetic chemicals, batteries, metal refineries, paints, corroded galvanized pipes, and petroleum products	Immunological and gastrointestinal systems, liver, testes, kidneys, brain, and bones.	3
Zn	Brass coatings, metal alloys, pigments, electroplating, industrial waste, aerosol deodorants, rubber products, some cosmetics, and coal combustion.	Vomiting, nausea, anemia, stomach cramps, convulsions, and skin irritation	3000

(*Continued*)

Table 2.2. (*Continued*)

HM	Human-induced sources	Key organs and effects	Maximum permitted level ($\mu g \ l^{-1}$)
Hg	Runoff from landfills and agriculture, combustion of fossil fuels, electrical appliances, mining, electronics and plastics industries, laboratory devices, refineries, electrolytic production of caustic soda and chlorine, and paper and pulp industries	Liver, immunological and reproductive systems, lungs, cardiovascular and endocrine systems, kidneys, and brain.	6
Ni	Petroleum refineries, production of stainless steel and nickel alloy, metal alloys, pulp and paper mills, electroplating industries, fertilizers, and battery plants	Pulmonary fibrosis, asthma, skin, gastrointestinal distress, kidneys, and lungs	70

reduced in bodies of water that may or may not come into contact with the ecosystem [52]. The various HMs, their permissible limits according to the EPA and the World Health Organization (WHO), and their effects on human organs [5, 48, 50] are depicted in table 2.2 and figure 2.13.

The range of successful traditional approaches to removing pollutants from water, which have already been positively verified, includes ion exchange, reverse osmosis, precipitation, ADPN, membrane separation, etc. ADPN is the most capable and frequently used approach owing to its ease of operation, high effectiveness, and commercial benefits. The ADPN method can be substantially improved by pioneering the use of external influences such as EFs, MFs, and irradiation. In WT, MFs have been utilized as a valuable tool due to their ease of access, effectiveness, capability, energy efficiency, and insignificant effect on the ecosystem [11].

2.6 Dyes

The excessive use of dyes in modern times has caused widespread concern due to the release of dye into the environment. It is projected that about 7 000 000 tons of dyes are utilized worldwide per year, and about 10–15% of the dyes released into the aquatic environment are released without being subjected to any controls [6].

As of 2021, the overall market for pigments and dyes was estimated to have a value of 36.4 billion US dollars. Dye toxicity is a huge concern for dye manufacturers, and they have enthusiastically explored approaches for avoiding dye release into the environment during the production of dye. Furthermore, dyes are used in huge quantities for various colors of paints, coatings, plastics, and textiles. Hence, the ecosystem is exposed to pollution by dyes from various manufacturing sources

[53]. Despite dye's undisputable importance, this industrial sector is unique in the overall picture of the key contaminants and consumes high volumes of chemicals and fuels. The vast utilization of water in different processes of its production chain, such as bleaching, dyeing, washing, etc. is considered critically important [54].

This is the key environmental impact of textile manufacturing, but the impacts ensuing from unprocessed runoff discharged into water resources generally make up 80% of the total discharge created by this industry. In the composition of most of the WWR produced by the fabric industry, there are comparatively elevated levels of chemical oxygen demand (COD) and biochemical oxygen demand (BOD). More importance should be ascribed to the huge quantity of non-decomposable carbon-based compounds, particularly textile dyes [54]. The volume of freshwater utilized by various countries in m^3 in the production of clothing or by textile industries is shown in figure 2.14 [55].

When dyes are dispersed in an aqueous medium, their effects depend on their FGs, color, and ionic charge. The last method of classifying dyes, which is very important, is based on the ionization of the dye molecules and has a substantial influence on the sorption performance of various NMs. Therefore, they are categorized as ionic (cationic (basic) and anionic (reactive, direct, and acidic)) and nonionic (vat and disperse) dyes, as depicted in table 2.3. The dyes cited in table 2.3 are generally carcinogenic and should be captured effectively in dye-processing plants to prevent them from being discharged into the environment [56, 57]. Unsurprisingly, all dyes are soluble, except for vat and disperse dyes. They can also contain trace elements such as Pb, Zn, Cu, Cr, and Co. Since their chemical structures are complex, artificial dyes are generally stable against oxidation agents, which leads to enhanced COD [57].

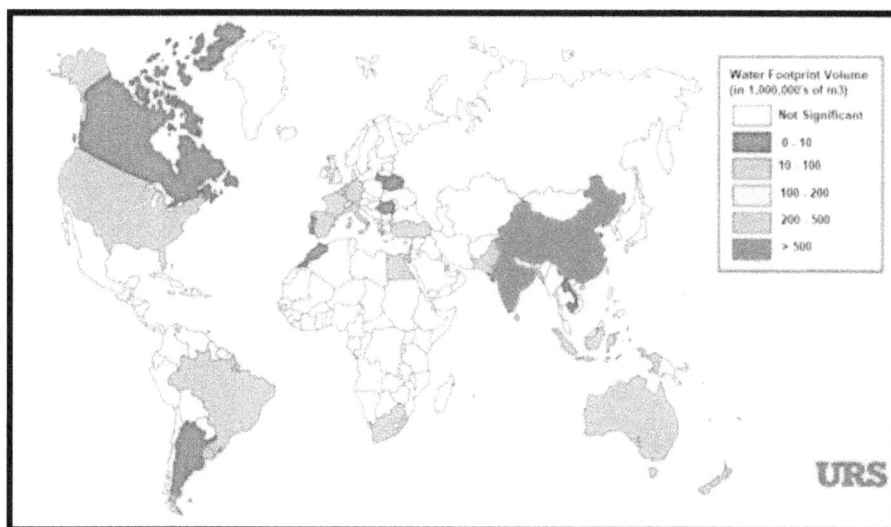

Figure 2.14. Water consumption by the textile industry in selected nations (m^3). Reprinted by permission from Springer Nature Customer Service Centre GmbH: [Springer Nature Link] [Environ Chem Lett] [55], Copyright (2019).

Table 2.3. Ionic and nonionic dye types, with their applications, properties, and toxicity.

Dye type	Examples	Properties	Solubility in water	Applications	Toxic effects
Acidic	Acid Red 183 (AR-183), Acid Orange 10 (AO-10), Acid Red 18 (AR-18), Sunset Yellow (SY), Acid Green 27 (AG-27), Methyl Orange (MO), Amido Black 10B (AB-10B), and Indigo Carmine (IC)	Anionic, WS	WS	Nylon, paper, wool, silk, inkjet printing, and leather	Carcinogenic
Cationic	Methyl Blue (MB), Basic Violet 10 (BV-10), Janus Green (JG), Basic Green 5 (BG-5), and Rhodamine 6G (R-6G)	WS, release of color cations	WS	Antiseptic for biomedicine, paper, treated polyesters, and nylon	Carcinogenic
Disperse	Disperse Orange 3 (DO-3), Disperse Red (DR), Disperse Red 1 (DR-1), and Disperse Yellow 1 (DY-1)	Nonionic for aqueous/HP dispersions, insoluble in water	Insoluble in water	Acrylic fibers, polyester, nylon, cellulose, and cellulose acetate	Skin allergenic and carcinogenic
Direct	Congo Red (CR), Direct Red 23 (DR-23), and Direct Blue 86 (DB-86)	Anionic, WS, and chelated with metal salts	WS	Cotton, paper, leather, and regenerated cellulose	Cancer of the bladder
Reactive	Reactive Black 5 (RB-5), Reactive Green 19 (RG-19), Reactive Blue 4 (RB-4), Reactive Red 195 (RR-195), Reactive Red 198 (RR-198), Reactive Red 120 (RR-120), Reactive Blue, and 19 (RB-19)	Produces brighter colors and a high wash grip owing to the covalent bond to the fibers	WS	Wool, cotton, inkjet printing of textiles, and nylon	Dermatitis, conjunctivitis, allergic reactions, rhinitis, and work-related asthma
Vat	Vat Blue 4, (VB-4), Vat Green 11 (VG-11), Vat Orange 15 (VO-15), Vat Orange 28 (VO-28), and Vat Yellow 20 (VY-20)	Insoluble in water and reduced to a weakly acidic leucon form by NaOH	Insoluble in water	Cellulose and fibers	Irritation of the skin and mucous membranes, burns.

The colors linked with fabric dyes not only cause harm to water resources but also prevent light from penetrating water, which leads to a decrease in the level of photosynthesis and dissolved oxygen, impacting the entire marine biota [58]. The direct discharge of unprocessed dyes contained in WWR into natural water resources has an antagonistic impact on photosynthesis in marine ecosystems. It has teratogenic or mutagenic effects on marine creatures and fish species due to the presence of metals and aromatics. Additionally, the presence of dyes in the ecosystem has insignificant to severe toxic impacts on human health, including mutagenic, allergic, and carcinogenic effects, dermatitis, and kidney disease [59]. The ingestion of dyes by humans can cause skin cancer, mental instability, and increased blood pressure. Dye pollution can also reduce the capacity of the blood to transmit active oxygen throughout the body. Therefore, the elimination of dyes from industrial runoff is vital [60]. Hence, it is essential to protect the ecosystem from unrestricted toxic dye runoff in water by exposing it to various chemical, physical, and biological treatments, or a combination of these [59].

2.7 Approaches for heavy metal and dye removal

2.7.1 Magnetic-field-assisted adsorption process

One of the four main elements in the earth's crust is Fe, which is considered an indispensable metal to human civilization and is consumed at a rate of around a million tons yearly. It is also a common metal that has a variety of uses, with recent applications in WWT in the form of Fe-based materials. This is due to their extraordinary reactivity, eco-friendly nature, low cost, magnetic features, and extraordinary sorption capacity (Q_e). Among the sixteen different forms of FeO phases created via various synthesis approaches and described in the research literature, hematite ($\alpha - Fe_2O_3$), magnetite (Fe_3O_4), and maghemite ($\gamma - Fe_2O_3$) are the most widespread Fe-based nanosorbents [27].

The ADPN approach involves an interaction between an adsorbate and NPs, which is dependent on their physicochemical properties such as particle size, surface charge, aggregation, SA, surface covering, morphology, porosity, etc. Also, the ability of the ADPN process to support recycling has shown its usefulness and makes the process economically practicable. The aggregation of NPs is impacted by their mechanical, magnetic, electrical, and optical properties, such as particle size, surface charge, and composition. NPs lacking magnetic properties have restricted application in water purification owing to the difficulty of separating them from aqueous solutions. The existing benefits of MNPs are due to their high SA, their size and shape, which rely on catalytic properties and can be extracted from solution using an MF [40, 49, 61].

MNPs have been progressively utilized to sequester harmful contaminants from water bodies owing to their excellent physicochemical features based on their size, morphology, and biocompatibility. Their magnetic properties also lead to their great reputation for sorbing various toxic contaminants from water [4]. MFs are common in nature and have been observed to affect the structure of water, as observed using its spectrum, thermodynamic features, etc [62].

The application of MFs has been known for centuries, with the induction idea presented as early as 1830 by Michael Faraday. This concept states that an electrical current is induced when an MF flux is traversed by a flow of ions or conductive material. While MF applications were swiftly pursued to verify Faraday's claim, global acceptance from academics and capitalists was missing. Yet, all electrical power systems comprising motors and generators are based on Faraday's results. For use in WT, Faraday's findings were improved by Faunce and Cabell in the design of electromagnetic equipment that was fitted to a cooler system to reprocess and treat hard water. Owing to the reputation of this equipment, MF use in this field has increased significantly. The first commercial magnetic equipment for WT was patented by Theo Vermeiren in Belgium. The first suggestion for the use of MF in WT was made by Russian scientists. It was later discovered that water and other fluids can be magnetically charged when they are placed in contact with a permanent magnet (PM) of suitable strength for a significant time [63].

Controversy has continued for over 50 years regarding the magnetic treatment approach for industrial and domestic water. This physical treatment approach helps to avoid the application of chemicals such as polyphosphates or corrosive substances that are rare and can be detrimental to human life as well as damaging to the ecosystem. The autonomous assessment of magnetic treatment performance has been highly contentious, while the increase in the number of viable magnetic treatment devices might seem to confirm the efficiency of MFs in water processing. Various assertions have been made that MFs alter water's physicochemical features by inducing surface tension and chemical equilibrium, nucleation and growth, and pH value [64].

Magnetism is a unique physical property that justifies its use in the purification of water by impacting the physical properties of contaminants in water. In combination with other procedures, it can enhance the effectiveness of decontamination technology. This physical treatment approach assists in avoiding the application of expensive chemicals that can be detrimental to human life and disruptive to the ecosystem. This technology is currently being applied using either PMs or high-gradient magnetic separation (HGMS) in combination with magnetic ADPN, magnetic seeding, or electromagnetic equipment [63].

Static magnetization combined with the ADPN method has attracted special attention because it is easy to implement, low cost, and eco-friendly [52]. MF improves the ADPN process in various ways by enhancing the orientation of the adsorbate molecule with respect to the pores of the adsorbent. An applied MF with a velocity (v) substantially impacts the zeta (ζ) potential and particle size distribution in the ADPN medium, which is ascribed to the Lorentz force (LF) ($\vec{F} = q \cdot \vec{v} \times \vec{B}$) exerted on the charged solid particles or moving ions in aqueous solutions. For charged particles (CPs), this force linearly increases their velocity and the orthogonal vector component of the MF strength. LF(s) are created when CPs move in the direction perpendicular to the MF direction in the same plane, with the LF acting in the z-plane. Subsequently, the particles' surface charges are displaced from their original positions, thereby instigating unstable particles. The unstable particles move

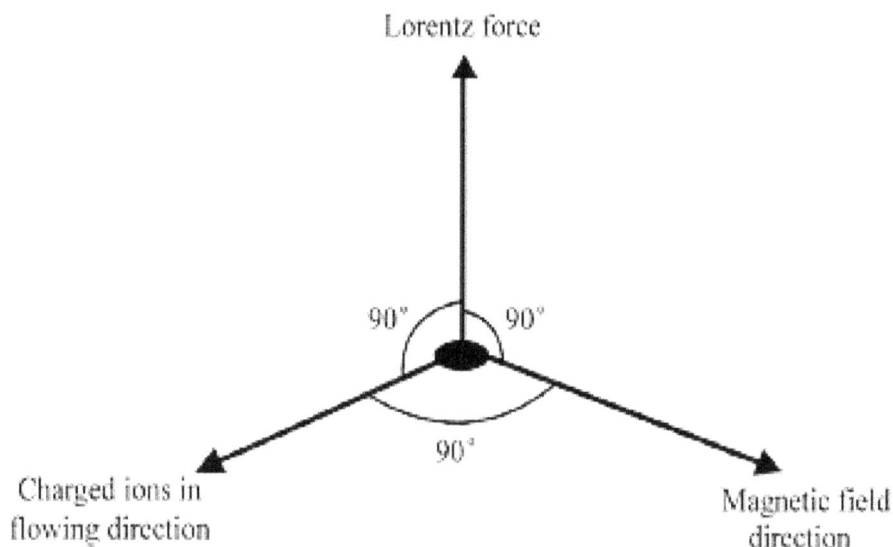

Figure 2.15. LF acting on moving charged ions or particles in an applied MF [63] [27 Jun 2012], reprinted by permission of the publisher (Taylor & Francis Ltd, https://www.tandfonline.com).

haphazardly, colliding with each other, thereby causing them to aggregate (figure 2.15). This improves linearly with the particles' charge and velocity and the orthogonal vector component of the MF strength, leading to improved interaction between the sorbent and adsorbate in the aqueous solution. In modest MFs of 0.01–0.1 T, the LF causes ions to change from a high-speed fluid layer to a low-speed layer, triggering ion strengthening near a solid boundary such as the catalyst surface. The higher the MF intensity, the greater the impact of the LF [9, 13, 63, 65].

Three design factors can be considered in the use of MFs for the supported ADPN process; these include the source type, the field line arrangement or trajectory, and the strength of the MF. MF sources are classified by their means of generation: either through direct current (DC), permanent magnets, or alternating current (AC) electromagnets. The latter contain a solenoid made from electrically conducting materials. The passage of electric current through the solenoid results in the creation of an MF in the electromagnet cavity. According to their applications, solenoid designs vary from annular to cylindrical and linear structures, while the strength of the MF created depends on the intensity, current potential, and number of turns [52]. In AC electromagnets, the current strength and direction vary in cyclic periods during the passage of electric current. This kind of electromagnet is employed to create MFs used in several ways, ranging from heating by induction to the removal of contaminants from WWR and rapid seed germination under magnetic pulse application. DC electromagnets are generally applied to assist in the process of removing contaminants from water and are wholly comparable to AC electromagnets. A regular and continuous field is generated solely by DC electromagnets due to the nature of their electric current, while AC electromagnets also create pulsations. The core elements used in the creation of permanent electromagnets are non-ceramic and ceramic ferromagnetic materials such as Ni, Co, Fe, silica, and rare metals. The coercive field and the field strength of the material are two factors that determine the magnetic force that these magnets produce. In the design of ADPN systems, the use of an MF is dependent on the MF orientation [52].

The performance of individual systems is directly influenced by the various techniques for magnetic applications. A PM generally generates a uniform MF, but this field can be varied widely by altering the arrangement and orientation of the magnet. Also, various PM shapes can produce diverse MFs. In cases where electromagnetic devices are utilized, a dynamic MF is usually obtained [63]. The magnetic exposure time and induction are given greater consideration as magnetic treatment conditions, while the fluid velocity has no significant influence on the magnetic treatment process [64].

The use of EMFs in WWR is promising due to their low cost, accessibility, eco-friendliness, low energy usage, excellent surface electronegativity, enhanced effectiveness, and ease of application. MFs can be stimulated to control the ADPN method, fast-track the stage-wise process of ADPN, or selectively enhance the sorption of specific pollutants via mass transfer. They are categorized as weak (<0.001 T), moderate (0.001–1 T), strong (1–5 T), and ultrastrong (>5 T) MFs. They can also be categorized as static or dynamic MFs based on their change over time. Importantly, the physical and chemical characteristics of sorbates that are generally considered are the hydration number, charge, valence, electronegativity, and ionic radius. These factors are considered in the selection of a sorbent material (SM) and determine the ADPN mechanism, since they impact mobility and forces [66].

The operational parameters of the MF determine the magnetic gradient (MG) size, thus impacting the MF. The greater the MG, the stronger the impact of the MF and the more noticeable the action of the magnetic phenomenon on ADPN. The three key types of MF are static MFs (SMFs), rotating MFs (RMFs), and pulsed MFs, as depicted in figure 2.16 [13, 67]. An SMF is a continuous MF where the magnetic flux density (MFD) at a given point in space does not fluctuate over time. Under an MF, the CPs are subjected to a force, and the impact of the MF is associated with the change in MFD at the given point. For an identical point in SMF space, the variation in the MFD is trivial; thus, the MG is minimal, and its influence is therefore minor. The two common ways to create a continuous MF are the application of DC to a wire to create an MF around the wire and the use of a PM to create an MF. In RMF generation, an MF with a continuous size and direction performs a constant rotational motion over time. At a specific point in the RMF, a sinusoidal MF strength can be observed that changes over time. The state of motion of the aforementioned particles and the transformation of the MF at this point are applied to determine the magnitude of the MG at this point. Under unvarying conditions at this point, the MF created by the RMF fluctuates more than an SMF; therefore, the RMF gradient acting at this point is huge, and the particles gain more MF energy that can overcome Brownian thermal motion between the molecules. This has more benefit in altering the orientation of the molecules. An RMF is created by a current flowing in a wire with continuous variation; alternatively, a PM can be mechanically driven to create an RMF [67].

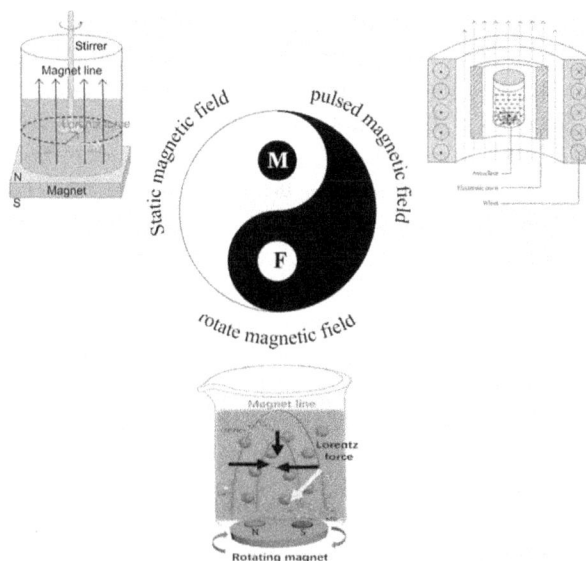

Figure 2.16. Various configurations for MF-assisted ADPN processes. Reprinted from [67], Copyright (2021), with permission from Elsevier.

In pulsed MFs (PMFs), the MFG changes continuously while the direction of the MF is constant. The PMF process utilizes an alternating oscillator to create alternating pulsed currents. By passing these currents into the coil of an electromagnet, PMFs of different shapes can be created. The structures of the PMF are such that the MF appears random. The varying frequency, waveform, and MF peak value can be adjusted as required, and the MF of the PMF varies more than the RMF, with the point of alignment transforming faster than the RMF and the degree of alignment changing significantly in a time on the order of milliseconds during the PMF application [67].

To improve the quality of WWT, MFs have generally been used in the ADPN separation approach, the photocatalysis procedure as a supplementary approach, and the biological degradation procedure. Recently, a technology that combines MNPs and MFs was used in the ADPN separation procedure and the biological degradation procedure to achieve the reuse of catalysts and adsorbents. Improved reuse efficiency was achievable due to the retrievability and the extraordinarily precise SA of MNPs. The impact of MF-assisted WWT technology is jointly controlled by the exposure time, field density (FD), and other experimental variables [13]. However, the use of MNPs in the presence of an alternating MF (AMF) produces heat. The heating process favors the ADPN of organic contaminants [68].

In a study by Rivera *et al* [68], electrochemically synthesized Fe_3O_4 NPs with a particle diameter of 12 nm were applied for the MF-assisted sequestration of Cr^{6+}. The optimum ADPN of Cr^{6+} was found to occur at a working pH of 3.5, a contact reaction time of 2 h, and a concentration of 2 g l^{-1} of synthesized Fe_3O_4 NPs. Increased removal of HM ions (Cr^{6+}) was observed in the presence of an EMF, which was considerably (30%) higher than the result obtained using a bath heating approach for Cr^{6+} removal. The sequestration of Cr^{6+} ions was determined to perfectly follow the pseudo second-order kinetic model (PSOKM). It was noted in Szatyłowicz and Skoczko's study that the applied MF positively influenced the sorption of Cu, Cd, and Pb ions by granular activated alumina (GAA), with an increased efficiency of 1.9%–8.2% reported for samples exposed to an MF. A removal rate of 95.4% was reported for this method of Pb ADPN by GAA, compared to the result obtained for a control sample processed without an applied MF (87.1%). For Cu ion removal, a 93.4% removal rate was reported using an MF, compared to the results obtained for control samples processed without an applied MF (91.5%). For Cd, a 77.7% removal rate was observed, in comparison to a Cd removal rate of 72.5% obtained without the application of an MF (figure 2.17). Based on the research findings, it was found that a combination of the MF and ADPN processes was effective in the remediation of these HMs. The influence of the MF on the ADPN of HMs by GAA was attributed to its ability to alter the structure of water and the aqueous solution. This resulted in changes in the viscosity of the solution, the surface tension, and the vaporization enthalpy, in turn leading to a variation in the sorption process of the solution components by GAA [69].

A magnetic propeller/agitator and reactor were used in a study by Fan et al to improve HM sequestration by zero-valent Fe (ZVFe). It was noted in this study that the application of a weakly generated MF considerably enhanced the removal of Cu^{2+}-–ethylene diamine tetraacetic acid (EDTA) by ZVFe from 10% to 98% after 2.5 h at a preliminary pH of 6. Based on kinetic studies, it was discovered that the pseudo first-order kinetic model (PFOKM) ideally characterized the MF-assisted ADPN of Cu^{2+} and As^{5+} by ZVFe, which resulted in improvements of 1.5–5.2 and 3.0–5.9 times, respectively. When the MF-assisted approach was used to treat industrial WWR containing Cu, and Zn, the sequestration of these HMs by ZVFe was significantly enhanced, meeting the industrial drainage standard after 20, 3, and

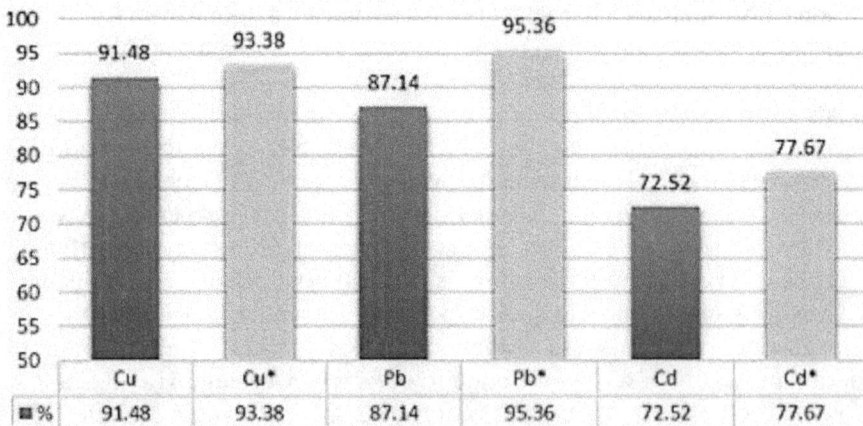

■ %	Cu	Cu*	Pb	Pb*	Cd	Cd*
	91.48	93.38	87.14	95.36	72.52	77.67

Figure 2.17. MF-assisted ADPN of Cu, Pb, and Cd ions by GAA. Reproduced from [69]. CC BY 4.0.

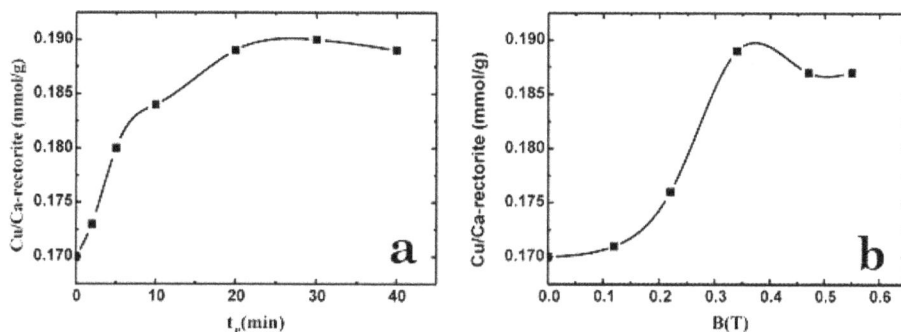

Figure 2.18. (a) Influence of MF exposure time and (b) MFD on Cu^{2+} elimination. Reprinted from [71], Copyright (2004), with permission from Elsevier.

25 min, respectively [70]. Rectorite clay samples saturated with Ca were used to synthesize a Ca–rectorite sorbent, which was employed for the sorption of Cu^{2+} under an SMF. The calculated area of the produced sorbent was 17 $m^2 g^{-1}$. Based on figure 2.18(a), it can be observed that the amount of Cu^{2+} sorbed by Ca–rectorite improved as the exposure time increased from 2 to 30 min, while the optimum amount of Cu^{2+} was reported to be sorbed at 30 min. There was also a sharp increase in the amount of Cu^{2+} sorbed by Ca–rectorite when the applied MFD was increased from 0 to 0.34 T, with 0.189 $mmol\,g^{-1}$ of Cu^{2+} sorbed at an MFD of 0.34 T (figure 2.18(b)). Based on the ζ potential results, it was noted that the ζ potential of the magnetized Ca–rectorite suspension in the absence of Cu^{2+} was significantly greater than that of the non-magnetized Ca–rectorite suspension under the same conditions. This was attributed to an increase in H^+ released from the amphoteric surface hydroxyl group of the SOH groups. Generally, the equilibrium constant was greater than that at equilibrium, leading to an increase in Cu^{2+} sorption by the sorbent; thus, MF treatment was seen to encourage equilibrium and the release of H^+ from the SOH groups. During Cu^{2+} sorption by a Ca–rectorite suspension under the influence of an MF, it was observed that the ζ potential of the suspension decreased as the colloid's diffusivity reduced. The influence of the MF was also generally attributed to the use of particles instead of the bulk solution [71].

This study described the impact of an externally generated field on the ADPN of Cr^{6+} using Fe_3O_4 NPs. A drastic increase in the percentage (%) of Cr^{6+} sequestered by the Fe_3O_4 NPs was noted as the external field intensity was increased from 0.5 to 3.0 A, with an elimination % of 98% observed at an electromagnetic current of 2.5 A and a fixed 10 V potential difference. This was attributed to the magnetization due to the superficial improvement of the magnetic sorbent's affinity for the HMs as the strength of the MF was increased until saturation was approached. The SP was excellent and well described by the Langmuir and Freundlich Isotherm models (LIM and FIM), with a maximum Q_e of 0.15 $mg\,g^{-1}$[72].

In Zn^{2+} adsorption by Na–rectorite (NR) under the influence of an SMF, it was observed that surface equilibrium for Zn^{2+} sorption to NR under the magnetic field was achieved in one hour. During exposure to the field, it was observed that the amount of Zn^{2+} sorbed by NR at an FD of 0.32 T and a concentration of $3 \ mmol \ l^{-1}$ increased as the field exposure time lengthened. The maximum percentage of Zn^{2+} sorbed by NR was observed at a field exposure time of 20 min and a concentration of $0.199 \ mmol \ g^{-1}$. It was also reported in this study that the amount of Zn^{2+} ions sorbed by NR at a given MFD at an MF exposure time of 20 min was improved, and the maximum percentage of Zn^{2+} sorbed was observed at 0.3 T. The ζ potential of the magnetized NR dispersion in the absence of Zn^{2+} was noted to be larger than that observed in the non-magnetized dispersed NR. Under the MF, the ζ potential was seen to reduce in the presence of Zn^{2+}[73].

Based on a state-of-the-art and noninvasive sorption-enhancing approach involving the use of SMFs, the sorption of a Cu, Ni, and Cd mixture was found to improve under the influence of an increased MF. The efficiency of the examined modification was found to be dependent on the concentration of the HMs at equilibrium, with an increase leading to a substantial decrease in the beneficial effect of the MF. The observed effect of the MF was greatly dependent on the metals being removed and their concentration. For the blend of HMs in the presence of the MF, it was found that the MF has a significant effect on the amount of Cu ions sequestered, while the removal of Cd and Ni was relatively poor under the effect of the MF. Based on a theoretical consideration of the impact of the EF on the SP at the interfaces of carbon nanotubes (CNTs), it was found that the structural changes to the highest occupied molecular orbital (HOMO) and lowest unoccupied molecular orbital (LUMO) orbitals of the sorbent particles resulted in an effective modification of the process. Based on a kinetic and isotherm study, it was found that the PFOKM, LIM, and the Langmuir–Freundlich model (LFM) best described the ADPN process in the presence and absence of an MF [74].

In a study by Zhao et al, the use of an MF was explored in the removal of nitrate $(NO_3^-{-}N)$ and ammonia $(NH_4^+{-}N)$ from water using 3D/2D $Mn_2O_3/g{-}C_3N_4$, leveraging their opposite charges. Extraordinary sequestration efficiencies of approximately 95% and 97% were reported for NO_3^-N and $NH_4^+{-}N$, respectively, under the influence of an MF. The benefits of this additional MF also included the provision of a driving force that caused NO_3^-N and $NH_4^+{-}N$ to approach the catalyst surface and enabled the separation of photoinduced charge carriers in this heterojunction by creating an LF [75].

A nonintrusive SMF was explored for the sorption of Cd^{2+} and Zn^{2+} ions by AC obtained from bituminous coal and bone char. It was noted in this study that, owing to the ferromagnetic property of AC produced from bituminous coal, it performed better in the ADPN of these HMs than AC obtained from bone char, which had PMC characteristics under the influence of an MF. For AC produced from bituminous coal, Q_e increases of 63% and 15% were measured for Cd^{2+} and Zn^{2+}, respectively [76].

A study by Aigbe et al examined the influence of a changing RMF in facilitating the elimination of fluoride (F^-) by prepared NCs (polypyrrole (ppy)/Fe_3O_4 NCs). The characteristic diffraction peaks obtained using XRD at 19°(111), 30°(220), 36° (311), 43°(400), 54°(422), 57°(511), 63°(440), 71°(442), 74°(533), and 75°(622) from the prepared NCs were attributed to the face-centered cubic Fe_3O_4 NP structure

(JCPDS no. 75-0449), while a peak at 25° was attributed to ppy. The average crystal size of the NCs based on the Debye–Scherrer formula (DSF) was 19 nm, in agreement with the average particle size determined using TEM (12 nm). Based on the research results, 78% of the F^- ions were sorbed by the NCs at a pH of 6 using a 0.019 T MF. Removal was also enhanced from 74%–78% as the MF exposure was increased over varying periods from 20 to 60 min. This was due to the aggregation of particles, which led to enhanced chain collisions and extensive particle contact with the solution containing F^-. This ensued from the gradient force applied to the particles as the MF was increased, resulting in better elimination of F^-. The process of F^- ADPN by the NC followed the FIM and the Temkin isotherm model (TIM) [11].

In another study by Aigbe *et al* (2018), an RMF was applied to stimulate a ppy/ Fe_3O_4 NC used for the elimination of Cr^{6+} ions. It was found that the MF influenced 99% of the Cr^{6+} ions sorbed to the NC, with optimum removal observed at a pH of 2 and an MF intensity of 19 mT. This was attributed to the increased interaction of particles with the Cr^{6+} solution ensuing from the improved velocity owing to the LF. This study found that the LIM and the linearized PSOM described the process of Cr^{6+} ADPN by the NC under the influence of an MF [9].

In a further study, Fe_3O_4 NPs were layered with ppy to create Fe_3O_4/ppy nanocomposites (NCs) using the coprecipitation approach. The prepared NCs were assessed for the effective elimination of CR dye under the effect of a changing MF generated by a three-phase induction motor. At an MF strength of approximately 19 mT, optimal removal of 94% of the CR dye was observed by the NCs at a pH of 4. As MF intensity increased (12–27 mT), an improvement in the percentage of CR dye eliminated was noted, which was attributed to increased CP velocity and repositioning. Based on the models studied, the LIM and the PSOKM perfectly described the ADPN of CR by the NCs under the influence of changing MFs. The optimum ADPN capacity (Q_m) based on the LIM was assessed to be 120 mg g^{-1}[77].

MF was also reported to be efficient in the treatment of surface water from a reservoir containing Fe, As, and Zn. It was found in this study that exposing the reservoir surface water to a 0.35 T MF for 30 min and using ferric chloride ($FeCl_3$) resulted in the elimination of 50 mg l^{-1} of Fe, HMs, and dissolved organic carbon (DOC). The MF was thought to improve particle interaction, leading to possible coagulation and aggregation of particles [78].

The impact of MF intensity on the elimination of Cr^{6+} ions by zero-valent Fe with various particle sizes was explored in a study to determine the comparative influences of the LF and a magnetic force gradient. It was observed that the elimination of Cr^{6+} ions was substantially improved by increasing the MF flux intensity from 0 to 2 mT. A further increase in the MF strength to 10 mT resulted in a slight enhancement in the amount of Cr^{6+} eliminated by the sorbent, signifying the negligible role of the LF. Additionally, there was improved removal of Cr^{6+} ions as the particle size was reduced under the influence of the MF, which was evident by the improved rate of Cr^{6+} elimination at a constant 2 mT MF compared to the case without an applied MF [79].

Fe_3O_4/ZnO NCs were used for the removal of P under the influence of a weak MF using the ADPN system. A significant amount of P was captured by the NCs under the influence of the weak MF, with a 97% removal of P at a pH of 7 and an intensity of 0.06 T. Based on the LF rule, the LF due to the weak MF applied to the P anions always pointed to the composite's center. The LF force blended with the electrostatic attraction accelerated the capture of P by the NCs' surface in a certain direction, thereby accomplishing the ultrafast capture of P. After the NCs were exposed to the weak MF, they exhibited better performance and magnetic memory. This magnetic memory reinforced the charges in the NCs, thereby improving the possible interactions with P. It was found that the pseudo second-order model (PSOM) described the sorption of P by Fe_3O_4/ZnO under the influence of the weak MF, which suggested a chemisorption process. It was also noted that the presence of other common ions did not impact the ADPN of P by Fe_3O_4/ZnO [80].

Synthesized carbonaceous sorbents with various magnetic properties (PMC, DMC, and ferromagnetic) were used for the capture of Pb^{2+} and F^- under an EMF. ADPN studies of the capture of Pb^{2+} confirmed an association between the sorption capacities and the SM's magnetic type. This behavior was not noted in the capture of F^- ions. This was attributed to the magnetic properties (χ) and ionic solvation (hydration number) of individual ions. The magnetic properties of the sorbates enabled them to create momentum, which acted either with or against the MF flow (LF). The MF was shown to impact ion hydration by improving the ions' magnetic properties in their hydrated state, providing a superior force that allowed them to overcome viscous resistance and improved the prospect of contact with the SM. Nonconformity in the charge trajectories was a function of the MF configuration, which created irregular movement, and this haphazard movement promoted contact with the SM and ions [81].

Zero-valent Fe FZD with Cu^{2+} and peroxymonosulfate was employed for the removal of Acid Orange 7 (AO7) under weak MFs. Based on the study results, 60% and 95% of the AO7 dye molecules were captured in the absence and presence of weak MFs, respectively, at a pH of 2 [82].

Lo *et al* (2021) studied the capture of Cu^{2+} ions by Fe_3O_4 NPs modified with a silica gel matrix to create Fe_3O_4/silica gel NCs under the influence of an MF. Under the effect of the MF, the percentage of Cu^{2+} ions captured by the NCs was further improved to 39% before ADPN. The magnetic domains were aligned with the NCs, resulting in an improved LF during the ADPN process. The process of Cu^{2+} ADPN by the NCs was described by the PSOM. This confirmed that the SP was governed by chemisorption, in which Fe^{2+} in the NCs acted as an electron donor to reduce Cu^{2+} ions [83].

Fe enclosed by montmorillonite, which was slowly added to a prepared water solution of clay (1 wt.%) at 35 °C, resulted in the preparation of magnetic clay (FeMAG) and pillared clay (FePILC). The prepared sorbent was assessed for the effective removal of MB under a pulsed MF. It was reported in this study that under the influence of a pulsed MF, there was increased elimination of MB dye. The LIM was also found to accurately describe the ADPN of MB by the various prepared materials under the influence of the pulsed MF. The optimal Q_m values obtained for FeMAG and FePILC were 835 and 223 mg g^{-1} under MF influence [84].

Hao *et al* (2012) studied the effect of an MF on the sorption of MB dye by organo-bentonite. The SA of the material, according to Brunauer-Emmett-Teller (BET) analysis, was 8.6 m^2 g^{-1} in a N$_2$ atmosphere. Efficient sorption of 100 mg l^{-1}

MB dye concentration by organo-bentonite at $1 \, \text{g} \, \text{l}^{-1}$ was reported at pH values of 7–8. With the application of the MF directly perpendicular to the bottom of the container, it was noted that the time required for the process to attain sorption equilibrium was faster compared to the case where the MF was applied parallel to the bottom of the container. Under the vertical orientation of the MF, the movement of dye molecules directly perpendicular to the magnetic lines of force resulted in the LF exciting the molecules of MB dye in the direction of the organo-bentonite surface, leading to improved removal of MB dye. In the parallel orientation of the MF, the bipolar attraction on both sides of the solution resulted in less sorption of MB dye in contrast to the bipolar mutual repulsion. Under MF exposure, the sorbate and sorbent properties are altered, and the sorbent surface morphology becomes less consistent, resulting in an increased organo-bentonite SA for improved sorption of MB dye. The organo- bentonite particles are impacted more by the MF due to the Fe_2O_3 constituent of the bentonite, which is sensitive to MFs. The FIM and the PSOKM data set were reported to best characterize the MB sorption by the SM. The Q_m values obtained by this study were 93 and 98 mg g^{-1} without and with field exposure, respectively [85].

In another study, the MF-assisted ADPN process was used in the capture of contaminants by zeolites and carbons. The results showed that under the influence of an MF, there was an increase in the amount of HMs and dyes captured, which was observed more with the zeolite than with the carbons. The capture of HMs was enhanced by using carbons under the influence of an MF, but these sorbents performed better in a binary solution treatment [86].

Sodium alginate/graphene/l-cysteine (SA/GR/l-cys) beads were explored for the effective capture of MB dye and Cu^{2+} under the effects of RMFs and SMFs. An improvement was observed as a result of using rotating and static MFs in the treatment of all modeled pollutants, with the RMF showing more than a 50% removal of all contaminants. Under the RMF effect, improved capture of various contaminants was attributed to the intense vibration of the MF-sensitive Fe_3O_4 NPs implanted in the sorbent matrix, which resulted in the creation of more wrinkles or sorption sites that improved the process. Also, the multidimensional LF created by the RMF enabled the movement of more pollutants from the solution to the active sites on the sorbent surface. The improved number of hydrogen bonds created under the RMF supported the ADPN process where there was a principal hydrogen bond interaction. The process was also impacted by the smaller ζ potential and enhanced chemical reaction observed under the influence of the RMF. The PSOM best fitted the sorption of these contaminants by the sorbent [87]. Overall, the studies reviewed in this chapter underscore the significance of MFs in improving the sorption of various pollutants found in WWR.

2.8 Magnetic-field-assisted adsorption mechanism

The key principle of magnetic applications is associated with the presence of molecular nucleation. As a physical approach, MFs have substantial physicochemical impacts owing to their magnetic interaction with electrons or atoms of WWR pollutants. Thus, they can considerably enhance the sequestration efficiency of WWR pollutants and provide a new method to reuse magnetic catalysts and sorbents during WWT procedures [13]. The broad concept and mechanism of effective MF applications are based on the MG, MF, magnetic memory, and LF. Molecules or particles are classified as positively charged (positive χ) or negatively charged (negative χ), where χ is the magnetic susceptibility. The magnetization (M) is given by $M = \chi_v H$ (where H is the applied MF expressed in electromagnetic units cm^{-3}, χ_v is the measured magnetic susceptibility of the molecules, and M is the particle magnetization after exposure to H). The M value of materials in the MF causes an alteration of its magnetoresistance (MR) (resistance variation of insoluble substances like contaminants, sorbents, and catalysts under the effect of MFs). In magnetic materials and non-magnetic materials, a greater MR is felt more in non-magnetic materials. Positive MR effects are related to the system of electrons in materials, where electrons interact through properties such as charge, spin, orbital degrees of freedom, and lattice vibrations. This impact occurs over a large range of temperatures (7×10^1–3.4×10^2 K) and MF intensities (zero to several T) and depends on the substance's spin-polarization properties and the LF created in the MF. Carriers have an increased collision probability under LFs created by fields of 0.05–1 T, accompanied by the subsequent production of an electron–hole recombination event, which is defined as a huge number of electrons falling into holes under the influence of the LF. Conversely, a negative MR is accompanied by decreasing resistance in insoluble substances, generally observed under modest MFs. This effect is triggered by the suppression of electron spin fluctuations associated with spin polarization involving the spin-charge direction of nuclei and electrons. Common materials used in WWT are sorbents and catalysts. The MR effects of these materials, particularly the negative MR effect, can encourage the migration of charge carriers such as Fe_2O_3 NPs (Fe^{3+}) toward energetic reaction sites; thus, more charge carriers can partake in the surface reaction, creating more energetic substances like .OH per unit time. Thus, WWT efficiency is improved, and the reordering of molecules eventually impacts WWT efficiency [13, 63].

Molecular substances are categorized as polar or nonpolar. In nonpolar molecules (NPMs), there is an overlap between the center of gravity of the positively charged nuclei and the electrons (the dipole moment is zero). In polar molecules (PMs), there is no overlap between the center of gravity of the positively charged nuclei and the electrons, as illustrated in figure 2.19. In the absence of an MF, PMs are randomly positioned. Hence, their negative and positive charges are difficult to bond with each other, even when an impact between molecules occurs. However, when the samples are subjected to an MF of a specific intensity, the PMs are effortlessly lined up in harmony with their positive and negative charges. In the absence of an MF, NPMs move constantly at random owing to the positive and negative charges, which coincide in the centers of molecules. This prevents coagulation, and under the influence of the MF, the positive and negative can be separated. The molecules are lined up in accordance with the MF direction, and with the ensuing alignment, these molecules are arranged methodically, initiating particle aggregation and coagulation. Additionally, the number of dipoles pointing in the

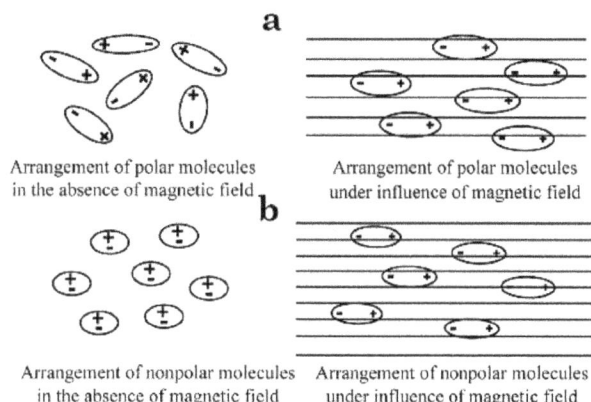

Figure 2.19. Influence of an MF on (a) a PM and (b) an NPM [63] [27 Jun 2012], reprinted by permission of the publisher (Taylor & Francis Ltd, https://www.tandfonline.com).

MF direction increases with the MF strength. This makes it more probable for the particles to aggregate or coalesce and for exceptional or superfluous particles or contaminants to be eliminated [13, 63].

MFs change the electrokinetic potential of CPs, thus altering molecular structure, polarization, and conductivity. Ultimately, the electrokinetic potential change created by MFs can lead to a reordering of CPs. The MFs' impact on PMC and DMC materials is generally caused by competition between electronic spin exchange, changes in spin direction, and the thermal movement of electrons facilitated by magnetic interactions. PMC molecules can be reordered in the direction of the MF. The reordering trend alters the intensity of the MF, which is reasonably weak in practical MFs. Moderately weak magnetic interactions are primarily observed in DMC states, and their interaction energy is trivial in relation to the thermal energy. Due to the spontaneous magnetism of ferromagnetic molecules in MFs, such molecules can be rearranged easily. The magnetic moment (μ-A.m^2) of each magnetic atom can be evaluated using equation (2.4). Due to the positive and negative charges of ferromagnetic molecules, they are easily separated owing to the difference in their magnetic moment under the influence of an MF. Hence, they can be effortlessly reordered despite their random arrangement in effluents:

$$\mu = \frac{V_m J_s}{N_A} \tag{2.4}$$

where V_m, J_s, and N_A are the ferromagnetic atomic volume (m^3 mol^{-1}), the saturated magnetic polarization (A m^{-1}), and Avogadro's constant [52, 63].

When the ADPN process is subjected to an MF, several factors and operating conditions rely on the magnetic and physicochemical nature of the sorbate and the sorbent, as well as the MF configuration and intensity. To elucidate the possible mechanism of the MF-assisted ADPN of various contaminants (in a sorbent–sorbate system), factors such as χ, the ionic radius, ion hydration, and the ζ potential are employed. The ζ potential is used to corroborate the influence of the MF on the sorbent–sorbate system. The ζ potential has been reported in different studies to be reduced or enhanced in some instances in sorbent–sorbate systems that were treated under the influence of an MF. Another possible explanation for the influence of MF exposure on the ADPN of various pollutants is that it alters the configuration of the water molecules, ions, and hydrated ions sorbed on the particles' surface. This leads to a denser, more efficiently sorbed layer, and the slipping plane may shift superficially away from the particle surface. This reduces the scale of the ζ potential, while the particles' superficial size becomes larger, decreasing their diffusivity [52, 66].

In understanding the selection of the SM and the ADPN mechanism, properties such as electronegativity, charge, valence, hydration number, and ionic radius are considered fundamental and assist in determining the forces and mobilities. Specifically, HMs exist in a positive ionic state; hence, their charge instability. However, when suspended in water, they remain unchanging in the solution, as they balance their charge with the electronic cloud of oxygen present in water molecules, which is highly negatively charged. Hence, the ions that are enclosed by the water molecules are pseudo-stabilized, and this is known as ion hydration. Ion hydration is important in the ADPN process, as it is recognized that individual ion ADPN forces represent the SM surface, which is applied to determine the benefit over other ions in actively occupied sites. This explains the Gouy–Chapman concept, which states that identical charges are sorbed with equal force on the sorbent surface. However, the ADPN strength is determined by the hydration ion, since those with a small hydrated radius are more strongly retained. This suggests an intensification in the ion size; hence, its reduced mobility. The viscous and electric forces impact the mobility of ions within a solution [52, 66]. The hydrated radius (r_h) and the hydration number (n_h) are defined by equations (2.5) and (2.6):

$$r_h = z_i e_o / 6\pi\eta u \tag{2.5}$$

$$n_h = \frac{r_h^3 - r_c^3}{r_{H_2O}^3} \tag{2.6}$$

where $z_i e_o$, η, u ($u = v/x$, $v =$ ion speed and $x =$ electric field), r_c^3, and $r_{H_2O}^3$ are the electric charge, viscosity, electric mobility due to the velocity ratio in the EF, crystallographic radius, and molecular radius of water [52, 66].

The hydration of ions is generally due to thermodynamic phenomena, and the variation in the hydration of ions attributed to the introduction of an MF has a greater or lesser impact depending on the hydration layer thickness of individual ions and the thermodynamic parameters of the hydration. Hence, it can be hypothesized that the MF favors ionic species' hydration owing to the creation of regular LFs, which assist in penetrating the boundary layer of the regular hydration, creating a molecular configuration that permits greater hydration. Additionally, sorbate transportation to the sorbent active sites is promoted by the impulse created by LFs and the electrostatic van der Waals forces, thereby realizing an advantage over the viscous force. Figure 2.20 elucidates the improvement in ionic hydration obtained under the influence of an MF, which is due to the better accommodation of

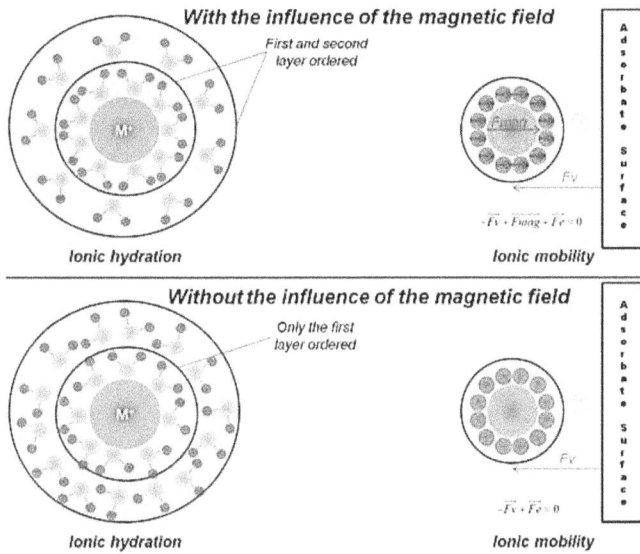

Figure 2.20. Contaminant removal mechanism in the presence and absence of an MF. Reproduced from [52]. CC BY 3.0.

water molecules surrounding the ions in the first and second layers of hydration, with less force felt in the second layer of hydration. The enhancement in the hydrated ions under the influence of an MF is due to the magnetic susceptibility of the water molecules and ions created by the magnetic force factor (F_{mag}), which is added to the electrostatic force (F_c), helping to overcome the viscous force (F_v) and increasing the probability with which ions strike the sorbent surface [52, 66].

The effective use of MFs strongly depends on the MF strength as well as the MG or the MF concentration, which varies regularly along magnetic devices. The energy (E) created by the magnetism of the material and the external magnetic field strength (H) for a volume (V) of the material is given by equation (2.7). It is assumed that the material load corresponds to the MF and density when M is unchanging. Hence, the magnetic interaction force (F) can be evaluated using equation (2.8), where X_o is the magnetic susceptibility of the material that contains the magnetized material. The critical factor that impacts the efficiency of MF use is dH/dx, which specifies the degree of variation of the MF strength with distance, and this is known as the MG. When the MF is unvarying, $dH/dx = 0$; under these conditions, the particles are magnetized and aligned with the MF. Hence, the particles are not exposed to a magnetic force (MAF) that would allow easy separation from the solution [63]:

$$E = -VM \cdot H = -V(Vx \cdot H) \cdot H \tag{2.7}$$

$$F = -(dE/dx) = (X_V - X_O) \cdot VH(dH/dx). \tag{2.8}$$

The MG is greater when the V of the material to be separated is trivial. Thus, the maximum MG is needed to create the strongest MAF on the particles for the purpose of separation. While in magnetic filtration, the MG is created to separate suspended solids from the solution. The MG is produced by permitting a solution comprising CPs to drift through coils that are magnetized by PMs situated outside the cylinders. The magnet's south and north poles are arranged alternately in opposite directions to produce an AMF. A high-intensity MG is created by the combination of the coil magnetization and the MF, which causes the flux lines to become very close to each other and more concentrated. As the coil density increases, the flux intensity improves. When the solution drifts through the coils, the CPs are attracted, leading to separation from the original solution. The MG intensity is greatly influenced by the strength of the MF and the coil magnetization characteristics [63].

2.9 Conclusions

The importance of sequestrating dissolved pollutants from industrial WWR is a key concern for global society, as these pollutants present a huge hazard to human health and the environment. An important and frequently employed approach for the elimination of these pollutants is the ADPN process, owing to its high efficiency, economic benefits, and ease of operation. There has been a global inclination toward the growing use of novel technologies such as MF-assisted ADPN processes using MNPs for the treatment of industrial WWR containing various contaminants. For acceptable results, the intensity and resolution of the MF and the fundamental magnetic properties of the sorbent and sorbate are pivotal. An intense MF is required due to the large number of existing spins that align with the MF according to the inherent nature of the MF, leading to sorbate motion and creating heterogeneity on the surface of the sorbent. In addition, the resolution of the MF controls the MF line direction. This alternative treatment process for various contaminants is important in today's global world owing to its ease of use, ecological purity, and safety.

References

[1] Lin L, Yang H and Xu X 2022 Effects of water pollution on human health and disease heterogeneity: a review *Front. Environ. Sci.* **10** 880246

[2] Ahmed M, Mavukkandy M, Giwa A, Elektorowicz M, Katsou E, Khelifi O, Naddeo V and Hasan S 2022 Recent developments in hazardous pollutants removal from WWR and water reuse within a circular economy *NPJ Clean Water* **5** 12

[3] Bayuo J, Rwiza M, Sillanpää M and Mtei K 2023 Removal of heavy metals from binary and multicomponent adsorption systems using various adsorbents–a systematic review *RSC Adv.* **13** 13052–93

[4] Prabhu S *et al* 2023 Magnetic nanostructured adsorbents for water treatment: structure-property relationships, chemistry of interactions, and lab-to-industry integration *Chem. Eng. J.* **468** 143474

[5] Aigbe U, Ukhurebor K, Onyancha R, Osibote O, Darmokoesoemo H and Kusuma H 2021 Fly ash-based adsorbent for adsorption of heavy metals and dyes from aqueous solution: a review *J. Mater. Res. Technol.* **14** 2751–74

[6] Aigbe U, Maluleke R, Lebepe T, Oluwafemi O and Osibote O 2023 Rhodamine 6G dye adsorption using magnetic nanoparticles synthesized with the support of vernonia amygdalina leaf extract (Bitter Leaf) *J. Inorg. Organomet. Polym. Mater.* **33** 1–20

[7] Ukhurebor K, Aigbe U, Onyancha R, Nwankwo W, Osibote O, Paumo H, Ama O, Adetunji C and Siloko I 2021 Effect of hexavalent chromium on the environment and removal techniques: a review *J. Environ. Manage.* **280** 111809

[8] Onyancha R, Aigbe U, Ukhurebor K and Muchiri P 2021 Facile synthesis and applications of carbon nanotubes in heavy-metal remediation and biomedical fields: a comprehensive review *J. Mol. Struct.* **1238** 130462

[9] Aigbe U, Das R, Ho W, Srinivasu V and Maity A 2018 A novel method for removal of Cr (VI) using polypyrrole magnetic nanocomposite in the presence of unsteady magnetic fields *Sep. Purif. Technol.* **194** 377–87

[10] Aigbe U, Ukhurebor K, Onyancha R, Osibote O, Kusuma H and Darmokoesoemo H 2022 Measuring the velocity profile of spinning particles and its impact on Cr (VI) sequestration *Chem. Eng. Process.-Process Intens.* **178** 109013

[11] Aigbe U, Onyancha R, Ukhurebor K and Obodo K 2020 Removal of fluoride ions using a polypyrrole magnetic nanocomposite influenced by a rotating magnetic field *RSC Adv.* **10** 595–609

[12] Tang S and Lo I 2013 Magnetic nanoparticles: essential factors for sustainable environmental applications *Water Res.* **47** 2613–32

[13] Wang Y, Gu X, Quan J, Xing G, Yang L, Zhao C, Wu P, Zhao F, Hu B and Hu Y 2021 Application of magnetic fields to WWR treatment and its mechanisms: a review *Sci. Total Environ.* **773** 145476

[14] Ali A, Shah T, Ullah R, Zhou P, Guo M, Ovais M, Tan Z and Rui Y 2021 Review on recent progress in magnetic nanoparticles: synthesis, characterization, and diverse applications *Front. Chem.* **9** 629054

[15] Stiufiuc G and Stiufiuc R 2024 Magnetic nanoparticles: synthesis, characterization, and their use in biomedical field *Appl. Sci.* **14** 1623

[16] Shukla S, Khan R and Daverey A 2021 Synthesis and characterization of magnetic nanoparticles, and their applications in WWR treatment: a review *Environ. Technol. Innov.* **24** 101924

[17] Ijaz I, Gilani E, Nazir A and Bukhari A 2020 Detail review on chemical, physical and green synthesis, classification, characterizations and applications of nanoparticles *Green Chem. Lett. Rev.* **13** 223–45

[18] Aigbe U and Osibote A 2024 Green synthesis of metal oxide nanoparticles, and their various applications *J. Hazard. Mater. Adv.* **13** 100401

[19] Liu Y, Bai J, Duan H and Yin X 2017 Static magnetic field-assisted synthesis of Fe3O4 nanoparticles and their adsorption of Mn (II) in aqueous solution *Chin. J. Chem. Eng.* **25** 32–6

[20] Kobylinska N, Klymchuk D, Khaynakova O, Duplij V and Matvieieva N 2022 Morphology-controlled green synthesis of magnetic nanoparticles using extracts of 'hairy' roots: environmental application and toxicity evaluation *Nanomaterials* **12** 4231

[21] Ahmed H, El-khateeb M, Sobhy N, Hefny M and Abdel-Haleem F 2023 Green synthesis of magnetite nanoparticles using waste natural materials and its application for WWR treatment *Environ. Sci. Proceed.* **25** 99

[22] Nyabadza A, McCarthy É, Makhesana M, Heidarinassab S, Plouze A, Vazquez M and Brabazon D 2023 A review of physical, chemical and biological synthesis methods of bimetallic nanoparticles and applications in sensing, water treatment, biomedicine, catalysis and hydrogen storage *Adv. Colloid Interface Sci.* **321** 103010

[23] Khan K, Rehman S, Rahman H and Khan Q 2014 Synthesis and application of magnetic nanoparticles *Nanomagnetism* ed J M G Estevez (London: One Central Press (OCP)) 135–9 http://www.onecentralpress.com/books/nanomagnetism-book/

[24] Aigbe U *et al* 2023 Utility of magnetic nanomaterials for theranostic nanomedicine *Magnetic Nanomaterials: Synthesis, Characterization and Applications* (Cham: Springer International Publishing) pp 47–86

[25] Jana S 2015 Advances in nanoscale alloys and intermetallics: low temperature solution chemistry synthesis and application in catalysis *Dalton Trans.* **44** 18692–717

[26] Majidi S, Zeinali Sehrig F, Farkhani S, Soleymani Goloujeh M and Akbarzadeh A 2016 Current methods for synthesis of magnetic nanoparticles *Artif. Cells, Nanomed., Biotechnol.* **44** 722–34

[27] Keshta B, Gemeay A, Sinha D, Elsharkawy S, Hassan F, Rai N and Arora C 2024 State of the art on the magnetic iron oxide nanoparticles: synthesis, functionalization, and applications in WWR treatment *Res. Chem.* **7** 101388

[28] Jang J and Lim H 2010 Characterization and analytical application of surface modified magnetic nanoparticles *Microchem. J.* **94** 148–58

[29] Gupta I, Sirohi S and Roy K 2023 Strategies for functionalization of magnetic nanoparticles for biomedical applications *Mater. Today Proc.* **72** 2757–67

[30] Wu W, He Q and Jiang C 2008 Magnetic iron oxide nanoparticles: synthesis and surface functionalization strategies *Nanoscale Res. Lett.* **3** 397–415

[31] Gao F 2019 An overview of surface-functionalized magnetic nanoparticles: preparation and application for WWR treatment *ChemistrySelect* **4** 6805–11

[32] Pinheiro P, Daniel-da-Silva A, Nogueira H and Trindade T 2018 Functionalized inorganic nanoparticles for magnetic separation and SERS detection of water pollutants *Eur. J. Inorg. Chem.* **2018** 3443–61

[33] Yaashikaa P and Kumar P 2022 Fabrication and characterization of magnetic nanomaterials for the removal of toxic pollutants from water environment: a review *Chemosphere* **303** 135067

[34] Faraji M, Yamini Y and Narges Salehi N 2001 Characterization of magnetic nanomaterials *Magnetic Nanomaterials in Analytical Chemistry* ed M Ahmadi, A Afkhami and T Madrakian (Elsevier) pp 39–60

[35] Khan S, Khan S, Khan L, Farooq A, Akhtar K and Asiri A 2018 Fourier transform infrared spectroscopy: fundamentals and application in functional groups and nanomaterials characterization *Handbook of Materials Characterization* (Cham: Springer) pp 317–44

[36] Cavalcante A, Dari D, da Silva Aires F, de Castro E, Dos Santos K, J and Dos Santos 2024 Advancements in enzyme immobilization on magnetic nanomaterials: toward sustainable industrial applications *RSC Adv.* **14** 179

[37] Asfaram A, Ghaedi M, Hajati S and Goudarzi A Synthesis of magnetic γ-Fe2O3-based nanomaterial for ultrasonic assisted dyes adsorption: modeling and optimization *Ultrason. Sonochem.* **32** 418–31

[38] Emegha J, Oliomogbe T, Ojo O, Muhammad F, Akpeji S and Ukhurebor K 2024 The types and characteristics of magnetic sorbents used for environmental sustainability *Environmental Applications of Magnetic Sorbents* (Bristol: IOP Publishing)

[39] Hajizadeh Z, Taheri-Ledari R and Asl F 2022 Identification and analytical methods *Heterogeneous Micro and Nanoscale Composites for the Catalysis of Organic Reactions* (Amsterdam: Elsevier) pp 33–51

[40] Almomani F, Bhosale R, Khraisheh M and Almomani T 2020 Heavy metal ions removal from industrial WWR using magnetic nanoparticles (MNP) *Appl. Surf. Sci.* **506** 144924

[41] Vohl S, Kristl M and Stergar J 2024 Harnessing magnetic nanoparticles for the effectiver-emoval of micro-and nanoplastics: a critical review *Nanomaterials* **14** 1179

[42] Scimeca M, Bischetti S, Lamsira H, Bonfiglio R and Bonanno E 2018 Energy Dispersive *X*-ray (EDX) microanalysis: a powerful tool in biomedical research and diagnosis *Eur. J. Histochem.: EJH* **62** 2841

[43] Faghihzadeh F, Anaya N, Schifman L and Oyanedel-Craver V 2016 Fourier transform infrared spectroscopy to assess molecular-level changes in microorganisms exposed to nanoparticles *Nanotechnol. Environ. Eng.* **1** 1–16

[44] Liu Z, Lei M, Zeng W, Li Y, Li B, Liu D and Liu C 2023 Synthesis of magnetic Fe3O4@ SiO2-(-NH2/-COOH) nanoparticles and their application for the removal of heavy metals from WWR *Ceram. Int.* **49** 20470–9

[45] Chen L, Gao Z, Li Q, Yan C, Zhang H, Li Y and Liu C 2024 A review: Comprehensive investigation on bandgap engineering under high pressure utilizing microscopic UV–Vis absorption spectroscopy *APL Mater.* **12** 030602

[46] Viter R and Iatsunskyi I 2019 Optical spectroscopy for characterization of metal oxide nanofibers *Handbook of Nanofibers* (Cham: Springer) pp 1–35

[47] Singh V, Ahmed G, Vedika S, Kumar P, Chaturvedi S, Rai S, Vamanu E and Kumar A 2024 Toxic heavy metal ions contamination in water and their sustainable reduction by eco-friendly methods: isotherms, thermodynamics and kinetics study *Sci. Rep.* **14** 7595

[48] Zamora-Ledezma C, Negrete-Bolagay D, Figueroa F, Zamora-Ledezma E, Ni M, Alexis F and Guerrero V 2021 Heavy metal water pollution: a fresh look about hazards, novel and conventional remediation methods *Environ. Technol. Innov.* **22** 101504

[49] Karnwal A and Malik T 2024 Nano-revolution in heavy metal removal: engineered nanomaterials for cleaner water *Front. Environ. Sci.* **12** 1393694

[50] Qasem N, Mohammed R and Lawal D 2021 Removal of heavy metal ions from wastewater: a comprehensive and critical review *Npj Clean Water* **4** 36

[51] Matei E, Predescu A, Predescu C, Sohaciu M, Berbecaru A and Covaliu C 2013 Characterization and application results of two magnetic nanomaterials *J. Environ. Qual.* **42** 129–36

[52] Virgen M, Vázquez O, Montoya V and Gómez R 2018 Removal of heavy metals using adsorption processes subject to an external magnetic field *Heavy Metals* (London: Intech Open)

[53] Teixeira R, Lima E, Benetti A, Naushad M, Thue P, Mello B, Dos Reis G, Rabiee N, Franco D and Seliem M 2023 Employ a Clay@ TMSPDETA hybrid material as an adsorbent to remove textile dyes from WWR effluents *Environ. Sci. Pollut. Res.* **30** 86010–24

[54] Lellis B, Fávaro-Polonio C, Pamphile J and Polonio J 2019 Effects of textile dyes on health and the environment and bioremediation potential of living organisms *Biotechnol. Res. Innov.* **3** 275–90

[55] Panda S, Aggarwal I, Kumar H, Prasad L, Kumar A, Sharma A, Vo D, Van Thuan D and Mishra V 2021 Magnetite nanoparticles as sorbents for dye removal: a review *Environ. Chem. Lett.* **19** 2487–525

[56] Aigbe U, Ukhurebor K, Onyancha R, Okundaye B, Pal K, Osibote O, Esiekpe E, Kusuma H and Darmokoesoemo H 2022 A facile review on the sorption of heavy metals and dyes using bionanocomposites *Adsorpt. Sci. Technol.* **2022** 8030175

[57] Elgarahy A, Elwakeel K, Mohammad S and Elshoubaky G 2021 A critical review of biosorption of dyes, heavy metals and metalloids from WWR as an efficient and green process *Clean. Eng. Technol.* **4** 100209

[58] Eldeeb T *et al* 2024 Biosorption of acid brown 14 dye to mandarin-CO-TETA derived from mandarin peels *Biomass Convers. Biorefin.* **14** 5053–73

[59] Dutta S, Gupta B, Srivastava S and Gupta A 2021 Recent advances on the removal of dyes from WWR using various adsorbents: a critical review *Mater. Adv.* **2** 4497–531

[60] Berenger K, Abollé A, Kouakou A, Gogbe V and Trokourey A 2023 Using modified activated carbon to remove methylene blue and rhodamine B from WWR *Am. J. Phys. Chem.* **12** 30–40

[61] Liosis C, Papadopoulou A, Karvelas E, Karakasidis T and Sarris I 2021 Heavy metal adsorption using magnetic nanoparticles for water purification: a critical review *Materials* **14** 7500

[62] Zhao L, Ma K and Yang Z 2015 Changes of water hydrogen bond network with different externalities *Int. J. Mol. Sci.* **16** 8454–89

[63] Zaidi N, Sohaili J, Muda K and Sillanpää M 2014 Magnetic field application and its potential in water and WWR treatment systems *Sep. Purif. Rev.* **43** 206–40

[64] Cai R, Yang H, He J and Zhu W 2009 The effects of magnetic fields on water molecular hydrogen bonds *J. Mol. Struct.* **938** 15–9

[65] Sivashankar R, Sathya A, Vasantharaj K and Sivasubramanian V 2014 Magnetic composite an environmental super adsorbent for dye sequestration—a review *Environ. Nanotechnol., Monitor. Manag.* **1** 36–49

[66] Onyancha R, Aigbe U, Ukhurebor K, Kusuma H, Darmokoesoemo H, Osibote O and Pal K 2022 Influence of magnetism-mediated potentialities of recyclable adsorbents for heavy metal ions removal from aqueous solutions–an organized review *Res. Chem.* **4** 100452

[67] Ren J, Zhu Z, Qiu Y, Yu F, Ma J and Zhao J 2021 Magnetic field assisted adsorption of pollutants from an aqueous solution: a review *J. Hazard. Mater.* **408** 124846

[68] Rivera F, Palomares F, Herrasti P and Mazario E 2019 Improvement in heavy metal removal from WWR using an external magnetic inductor *Nanomaterials* **9** 1508

[69] Szatyłowicz E and Skoczko I 2018 The use of activated alumina and magnetic field for the removal heavy metals from water *J. Ecol. Eng.* **19** 61–7

[70] Fan P, Jiang X, Qiao J and Li L 2018 Enhanced removal of heavy metals by zerovalent iron in designed magnetic reactors *Environ. Technol.* **39** 2542–50

[71] Zhang G, Yang X, Liu Y, Jia Y, Yu G and Ouyang S 2004 Copper (II) adsorption on Ca-rectorite, and effect of static magnetic field on the adsorption *J. Colloid Interface Sci.* **278** 265–9

[72] Muza L, Dube D, Ochieng A and Chiririwa H 2017 Investigation of the electromagnetic enhancement for the abatement of hexavalent chromium using magnetite as adsorbent *Iran. J. Sci. Technol., Trans. A: Sci.* **41** 859–65

[73] Zhang G, Liu Y, Xie Y, Yang X, Hu B, Ouyang S, Liu H and Wang H 2005 Zinc adsorption on Na-rectorite and effect of static magnetic field on the adsorption *Appl. Clay Sci.* **29** 15–21

[74] Rajczykowski K and Loska K 2018 Stimulation of heavy metal adsorption process by using a strong magnetic field *Water Air Soil Pollut.* **229** 1–7

[75] Zhao J, Li N, Yu R, Zhao Z and Nan J 2018 Magnetic field enhanced denitrification in nitrate and ammonia contaminated water under 3D/2D Mn_2O_3/g-C_3N_4 photocatalysis *Chem. Eng. J.* **349** 530–8

[76] González Vázquez O, Moreno Virgen M, Hernandez Montoya V, Tovar Gomez R, Alcantara Flores J, Pérez Cruz M and Montes Morán M 2016 Adsorption of heavy metals in the presence of a magnetic field on adsorbents with different magnetic properties *Ind. Eng. Chem. Res.* **55** 9323–31

[77] Aigbe U, Khenfouch M, Ho W, Maity A, Vallabhapurapu V and Hemmaragala N 2018 Congo red dye removal under the influence of rotating magnetic field by polypyrrole magnetic nanocomposite *Desalin. Water Treat.* **131** 328–42

[78] Duangduen C, Nathaporn A and Kitiphatmontree M 2006 The effects of magnetic field on the removal of organic compounds and metals by coagulation and flocculation *Phys. Status Solidi* C **3** 3201–5

[79] Li J, Qin H, Zhang W, Shi Z, Zhao D and Guan X 2017 Enhanced Cr (VI) removal by zero-valent iron coupled with weak magnetic field: role of magnetic gradient force *Sep. Purif. Technol.* **176** 40–7

[80] Li N, Tian Y, Zhao J, Zhan W, Du J, Kong L, Zhang J and Zuo W 2018 Ultrafast selective capture of phosphorus from sewage by 3D $Fe_3O_4@$ ZnO via weak magnetic field enhanced adsorption *Chem. Eng. J.* **341** 289–97

[81] González Vázquez O, Moreno Virgen M, Esparza González M, Hernández Montoya V, Tovar-Gómez R and Durán Valle C 2020 Analysis of the effect of a magnetic field applied to a process of adsorption of water contaminants using adsorbents of different magnetic orderings *Ind. Eng. Chem. Res.* **59** 13820–30

[82] Wang M, Zhang J, Zhao H, Deng W, Lu J and Ye Q 2019 Enhancement of oxidation capacity of $ZVI/Cu^{2+}/PMS$ systems by weak magnetic fields *Desalin. Water Treat* **161** 260–8

[83] Lo F, Kow K, Kung F, Ahamed F, Kiew P, Yeap S, Chua H, Chan C, Yusoff R and Ho Y 2021 Effect of magnetic field on nano-magnetite composite exhibits in ion- adsorption *Sci. Total Environ.* **780** 146337

[84] Tireli A, Marcos F, Oliveira L, do Rosário Guimarães I, Guerreiro M and Silva J 2014 Influence of magnetic field on the adsorption of organic compound by clays modified with iron *Appl. Clay Sci.* **97** 1–7

[85] Hao X, Liu H, Zhang G, Zou H, Zhang Y, Zhou M and Gu Y 2012 Magnetic field assisted adsorption of methyl blue onto organo-bentonite *Appl. Clay Sci.* **55** 177–80

[86] López S, Virgen M, Montoya V, Morán M, Gómez R, Vázquez N, Cruz M and González M 2018 Effect of an external magnetic field applied in batch adsorption systems: removal of dyes and heavy metals in binary solutions *J. Mol. Liq.* **269** 450–60

[87] Ma J, Ma Y, Yu F and Dai X 2018 Rotating magnetic field-assisted adsorption mechanism of pollutants on mechanically strong sodium alginate/graphene/L-cysteine beads in batch and fixed-bed column systems *Environ. Sci. Technol.* **52** 13925–34

IOP Publishing

Environmental Applications of Magnetic Nanomaterials
Under the Influence of Magnetic Fields

Uyiosa Osagie Aigbe and Kingsley Eghonghon Ukhurebor

Chapter 3

Magnetic resonance imaging using magnetic nanomaterials

Magnetic resonance imaging (MRI) is a non-intrusive and non-damaging approach that can provide three-dimensional (3D) images of organisms. The use of magnetic nanoparticles (MNPs) as contrast agents (CATs) permits medical scientists and specialists to substantially improve MRI specificity and sensitivity, as the CATs alter the fundamental features of the tissues inside an organism and thereby improve the existing image data. This chapter discusses the various synthesis and functionalization (FUCTN) approaches used for MNPs, their mechanism of action in MRI, and their use as MNP CATs. These CATs enable the imaging of various biological phenomena, providing insight into and detecting disease in organisms.

3.1 Introduction

Nanotechnology (NT) and nanoscience (NS) have greatly impacted studies in the fields of biology and medicine in recent years. Engineered nanoparticle (NP) applications are associated with fields such as drug targeting, gene delivery, and agents that improve the diagnostic viability of MRI and the development of unique imaging approaches. NPs possess novel properties such as extraordinary surface area (SA) to volume ratios, quantum attributes, and the ability to transport other compounds owing to their minute size. These features make NPs attractive for various medical applications [1].

In the past, scientific advancements in chemistry, medicine, physics, and engineering have led to biomedical imaging approaches with superior resolution and sensitivity, which help provide insights into biological phenomena and disease detection [2]. An accurate diagnosis is the first stage in recognizing a disease and articulating a therapeutic approach. The well-known and dependable imaging modalities that have been supporting clinicians in identifying an extensive range

doi:10.1088/978-0-7503-6377-8ch3

of diseases are ultrasonography, computed tomography (CTY), x-ray scans (XRSs), positron emission tomography (PETY), MRI, single-photon emission computed tomography (SPECTY), two-photon excited fluorescence (TPEF), and upconversion luminescence (UL) (figure 3.1). Advancing the performance of these techniques is a continuously growing field that has displayed constant development [3, 4].

Many of these approaches, such as XR, PETY, and SPECTY, require the use of detrimental ionizing radiation over extended exposures, which may lead to subsidiary problems [3]. These imaging approaches offer an array of benefits and limitations, depending on the specific medical requirements. MRI avoids these undesirable side effects, as it exploits external magnetic fields (MFs) created by the MRI machine. In addition, MRI with CATs surpasses in providing high penetration and three-dimensional (3D) precision [3, 5].

The merger of NT with medicine and molecular biology has led to the dynamic advancement of the contemporary field of bionanotechnology, which offers the exciting prospect of realizing unique materials, procedures, and phenomena. In 1950, Gilchrist *et al* treated lymphatic nodes and metastases (MeTas) by injecting metallic particles heated by an MF. Today, magnetic nanoparticles (MNPs) are innovatively used in the delivery of drugs, immobilization of enzymes, and a plethora of exciting biotechnological uses. Their remarkable features, such as their extraordinary SA, biocompatibility, uniform particle size, adsorption kinetics, magnetic moment, and superparamagnetism (SPM), can be tailored to specific uses during the MNP production process [6]. Through the advancement of highly focussed and effective CATs, MRI has developed into a multipurpose approach with various functions and an authoritative noninvasive (NI) imaging method in the

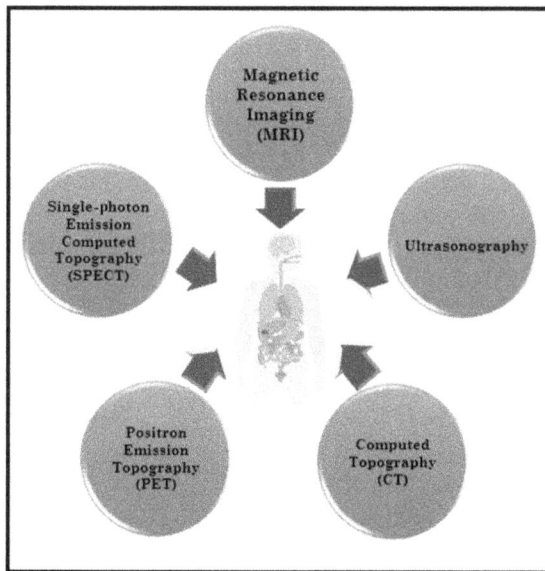

Figure 3.1. Various imaging modalities.

biomedical field. Its high resolution and exceptional soft tissue imaging are its key advantages over other *in vivo* (*InVi*) imaging approaches. MRI depends on huge MFs and radio frequencies (RFs), which make use of the relaxation times (RTs) of protons in water, lipids, and proteins—mobile molecules that exist in organs at various concentrations. It can create high-resolution (HR) soft tissue structural images with satisfactory deep tissue contrast [7].

This class of nanomaterials (NMs) is composed of metals with paramagnetic, ferromagnetic or superparamagnetic characteristics. This group comprises materials including transition metals (TMs) such as nickel (Ni), cobalt (Co), and iron (Fe) and metal oxides such as Fe_3O_4 and $\gamma - Fe_2O_3$. However, the pure forms of these metal NPs are chemically unstable in air and readily oxidized; hence, they are extraordinarily toxic. In contrast, their oxides are less prone to oxidation and have a consistent magnetic response. One group of frequently researched NPs with magnetic features is the superparamagnetic Fe oxide (FeO) NPs (SFeO NPs). These SFeO NPs exhibit novel features such as reduced Curie temperature, SPM, extraordinary magnetic susceptibility (χ), and coercivity. Most important are the properties of the particles that impact biodistribution inside the body, such as charge surface, particle size, and hydrophilicity/hydrophobicity. SFeO NPs have no net MM before being placed in an external MF (EMF), in contrast to their bulk ferromagnetic counterparts [8, 9].

MNPs are nanoscale materials that can be steered via an EMF owing to their ferromagnetic, superparamagnetic, or ferrimagnetic properties, which may offer features for biomedical uses. Their superparamagnetic properties are of remarkable importance owing to their strong magnetic interactions under an EMF, which disappear once the EMF is removed. These features enable ferrofluid design, as MNPs can be stabilized in solution, since no magnetic interactions take place when the EMF is removed. This leads to their *InVi* performance for cell marking, heat creation in hyperthermia treatment, drug systems steered by an MF, and imaging CATs. MNPs are generally composed of a core and a shell layer. An overall quantum effect is introduced by the core, which generally combines magnetic elements such as Fe, Co, and Ni with their corresponding oxides [10].

MNPs have gained substantial acceptance as MRI CATs owing to their unique biocompatibility, magnetic properties (MPs), and ease of surface FUCTN. Their surface, size, composition, and morphology can be altered remarkably, permitting baseline and heightened relaxivity (RELVTY). Furthermore, MNPs also offer a substantial advantage by effectively engaging with large biological structures, such as proteins, at the nanoscale level. In addition, their SA-to-volume ratio enables adaptability for specific purposes, including biochemical targeting strategies that depend on chemical coupling and stimulus-responsive designs based on encapsulation and a triggering mechanism [11, 12].

NP-based platforms are employed in MRI, and they have been shown to produce efficient results in accurate initial diagnosis and treatment due to their unique features, such as 3D resolution, strong soft tissue contrast, and nonexistent risk of radiation. These multifunctional imaging nanoprobes play a central role in improving the performance of their various modalities. They can be modified and concurrently employed in the remedial treatment of diseases such as diabetes, arthritis, cancer, neuropathic disorders, etc [2, 3]. They can also be used for tissue engineering.

In its basic form, a biomedical MNP platform is made up of inert core NPs and a biocompatible surface covering that offers stability under physical conditions. Furthermore, the integration of functional ligands is facilitated by the use of suitable surface chemistry. This integrated strategy allows MNPs to concurrently perform various functions, such as multimodal imaging, the delivery and real-time detection of drugs, and therapeutic methods. An important promising use of MNPs in nanomedicine is their ability to improve specific tissue proton relaxation and serve as an MRI CAT. A substantial challenge linked with the use of these MNP systems is their *In Vi* behavior. The efficiency of these various systems is frequently degraded owing to their identification and removal by the reticuloendothelial system (RES) before arriving at the target tissue and their inability to pass biological barriers like the blood–brain barrier or the vascular endothelium [13, 14].

For disease analysis or drug delivery, the physicochemical properties of FeO NPs, such as charge, surface modification, coating, size, shape, drug loading, FUCTN, preparation, and targeting approach, are considered. The effects of these NPs upon arterial administration are extremely reliant on the NPs' morphology, size, surface chemistry, and charge. These physicochemical features of NPs strongly impact their resulting biodistribution and pharmacokinetics. To improve MNP efficiency, various approaches, including size reduction and the attachment of antifouling polymers, have been utilized to conceal them. NPs applied in the biomedical field are required to be chemically stable, size standardized, and well dispersed in a liquid medium [14, 15].

For medical diagnosis and the development of targeted therapies, *In Vi* imaging is crucial. The key barrier to utilizing NPs is their toxicity. In the context of intravenous NP injection, the ideal formulation should show significant renal clearance, resulting from the absence of uptake by the RES and reduced material toxicity. The existing generations of hybrid NMs are extremely capable in terms of obtaining active images at the organ or cellular level [4]. In targeted delivery, physicochemical features are generally important in determining efficiency and accuracy. The two primary barriers that MNPs need to overcome to reach their target are the cellular and physiological barriers [16]. This chapter covers the basic principles of MRI as well as a synopsis of its current uses in medical practice.

3.2 Approaches for MNP synthesis

Diverse approaches to producing MNPs have been realized in recent years. The definitive goal of these approaches is to advance comprehensive control over MNP properties such as shape, saturation magnetization (M_s), MPs, size, and stability. Their compatibility with living creatures is a crucial matter that must be addressed, particularly in the case of their biological applications. An extensive array of synthesis approaches has been established over time. These synthesis approaches are grouped into chemical, physical, and biological methods. Individually, these approaches have discrete mechanisms and conditions for producing MNPs that are tailored to meet various applications and requirements [13, 17].

The properties of FeO NPs greatly rely on their shape, size, and the 3D distribution of the crystals inside the particles. To obtain the anticipated settings, a variety of the most suitable synthesis approaches should be used. The two key methods employed in MNP production are the bottom-up (BUP) and top-down (TD) approaches. In the key TD methods, the original bulk metal material is broken down to the nanomaterial (NM) level, thus creating NPs. In the BUP method, base Fe_xO_y molecules are precipitated and undergo the nucleation and growth stages until the NPs are obtained. FeO NPs can be produced via biological, physical, and chemical approaches, and the approach adopted should be based on achieving a low cost of NP production, high yields, and the ability to functionalize the NPs [3, 18, 19].

The most widely applied approach for the production of MNPs is the hydrolytic coprecipitation approach because the required reaction settings are readily scalable to an industrial scale due to the use of a stable colloidal NP suspension. In this approach, reducing agents combined with suitable surfactants are swiftly blended with a water-soluble medium of metallic precursors like Fe salts. Different types of MNPs can be produced using this approach. However, the drawbacks of this method are the uncontrollable characteristics of the NPs created, which have nonstandard shapes and sizes and reduced levels of crystallinity [20]. These inconsistencies hinder the use of NPs in the challenging assessment of and research into nanoscale magnetism and its optimized uses. Thus, most MNPs applied in systematic studies are made using the non-hydrolytic thermal decomposition approach. In this approach, a metallic precursor solution in a water-free organic solvent comprising suitable surfactants is heated to cause the thermal decomposition of its components, thereby forming crystalline NPs. This approach is readily controlled by regulating the time, precursor, surfactant type, and temperature. Accordingly, it is used to create outstandingly shaped, sized, and composition-tuned materials. This approach can also be utilized to create single-crystal NPs, which is a vital feature for their MPs [20, 21].

Artificial approaches currently applied for FeO NPs comprise the hydrothermal method, which characteristically results in mesoporous NPs, in contrast to the conventional coprecipitation method, which has several drawbacks including reproducibility problems related to morphological parameters (which have an inevitable impact on MPs) and the well-known sol–gel and auto-combustion approaches. Other artificial approaches highlight the use of natural extracts as bioinspired methods for creating MNP-based FeO NPs, with promising therapeutic possibilities. As reported by some studies [21–23], the various approaches employed for the synthesis of MNPs are displayed in figure 3.2.

3.3 MNP functionalization

An essential constituent of all MNP platforms for biomedical use is their surface coatings, owing to their tendency to aggregate due to their extraordinary surface energy. Colloidal electrostatic stabilization resulting from surface charge repulsion

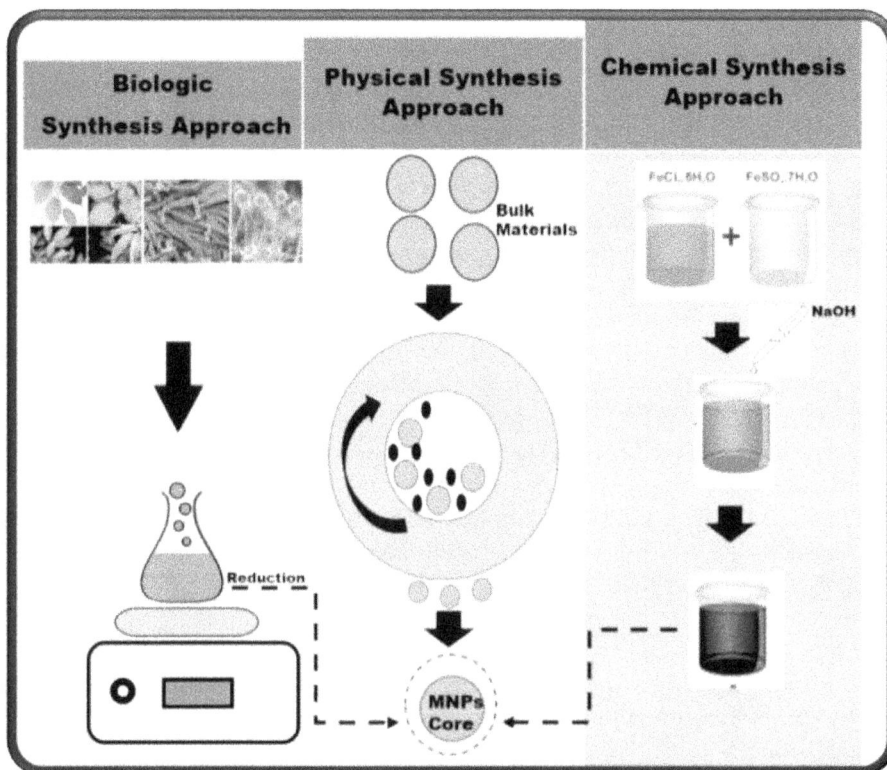

Figure 3.2. Synthetic approaches for MNP production.

on the NPs is generally unacceptable, as it causes buildup in biological solutions due to the existence of salts or other electrolytes that may counterbalance this charge [14]. For MNPs, there is an inescapable issue linked with their characteristic volatility over a long period, which is apparent in their loss of dispersibility, magnetism, and biocompatibility. Thus, creating an appropriate strategy is essential to protect MNPs from chemical breakdown. Meanwhile, biomedical applications of MNPs depend on various factors related to biocompatible MNPs' size, shape, magnetism, the physicochemical features of drug-laden MNPs, the geometry and strength of the field, the flow rate of blood, etc. MNPs functionalized with organic compounds during their synthesis are biodegradable and biocompatible while fully retaining their MPs. They are often used in the MRI, magnetic recording, electromagnetic shielding, and biological fields for magnetic cell separation, targeted drug delivery, etc [4].

To enhance the buildup of SFeO NPs in tumors, a promising approach is to combine the NPs with a targeting capability. In principle, this FUCTN would permit the imaging of FeO NPs by MRI and also offers the prospect of imaging cellular and subcellular functions and processes in living organisms without the risk of disturbing them, giving rise to so-called molecular MRI (mMRI) [13].

The useful application of such NPs in the medical and biological fields requires their surfaces to receive complex functionalization to increase the magnetic controllability of the MNPs, shield them from aggregation (thus leading to improved colloidal stability), and improve their surface stability and hydrophilicity. Furthermore, these carbon-based compounds can produce reactive functional groups (FGs) such as amine, carboxyl, aldehyde, and hydroxyl groups, which link to energetic sites of biological materials like DNA sequences and targeting ligands, proteins, enzymes, antibodies, etc. and promote the MNPs' non-toxicity and biocompatibility [4].

These additions broaden their biological application scope. In addition, to enhance their stability and thus avoid the aggregation of MNPs with elevated biocompatibility, they are covered with different organic compounds such as dextran, chitosan, poly(D, L-lactide) (PLA), gelatin, polyethylenimine (PEI), poly (methyl methacrylate) (PMMA), starch, alginate, and poly(ethylene glycol) (PEG), i.e. particularly hydrophilic organic compounds. Additionally, MNP surface FUCTN enhances their pharmacokinetics, personalized drug loading and discharge characteristics, and endosomal release. Functionalized MNPs offer the promise of transporting other energetic targeting moieties, imaging agents, and drugs through physical interactions and covalent linkages [4, 24, 25]. Multifunctional capabilities can be provided by a single type of particle or a blend of two or more types of NPs that have more than one function. A single type of NP can be multiply functionalized by combining core and shell types, using a multicomponent core, combining a core with different shell types, or asymmetrizing a core or a shell [4].

The approaches for MNP surface FUCTN include *in situ* (IS) and postsynthesis FUCTN approaches. The IS surface FUCTN approach is a single-pot synthesis approach that includes FUCTN, yielding a narrower particle size distribution. The reaction mixture is simultaneously fed with the MNP precursor and the coating material. The coating process starts as soon as nucleation occurs, hence preventing further particle growth. By selecting the appropriate stabilizing agent, NPs can be stabilized and protected from aggregation using IS FUCTN [24, 26].

A more popular coating approach with an increased scope for MNP FUCTN is the postsynthesis FUCTN approach. This approach toward FUCTN involves the production of MNPs and the modification of their surfaces. Bifunctional compounds are employed for this, in which a chelating binding group is primarily reacted, followed by a change of the linking site group to the definitive FG. In this approach, the synthesized MNP surfaces are covered with various functional materials using chemical or physical mechanisms such as hydrogen bonds, covalent bonds, electrostatic forces, affinity interactions, etc [4].

As reported by some studies [4, 24], the mechanism of the surface FUCTN of MNPs is based on ligand exchange, ligand addition, or ligand encapsulation. Ligand exchange is the replacement of hydrophobic ligands (HDLs) with hydrophilic ligands (HCLs) that have stronger binding properties. Ligand addition consists of the accumulation of a ligand on the MNP surface without a capping agent, the development of an inert layer on the surface of the MNP using ionic or an unknown interaction, covalent bond formation between the original and existing ligands, or the addition of hydrophobic species to the hydrocarbon shells of MNPs covered with ligands. In ligand encapsulation, HDLs on MNPs are coated with amphiphilic substances, which is achieved by adding the hydrophilic constituent of the materials to the solution and inserting the hydrophobic constituent of the material with the key ligand on the MNP surface. See figure 3.3(a–c), as adapted from Gupta *et al* [4].

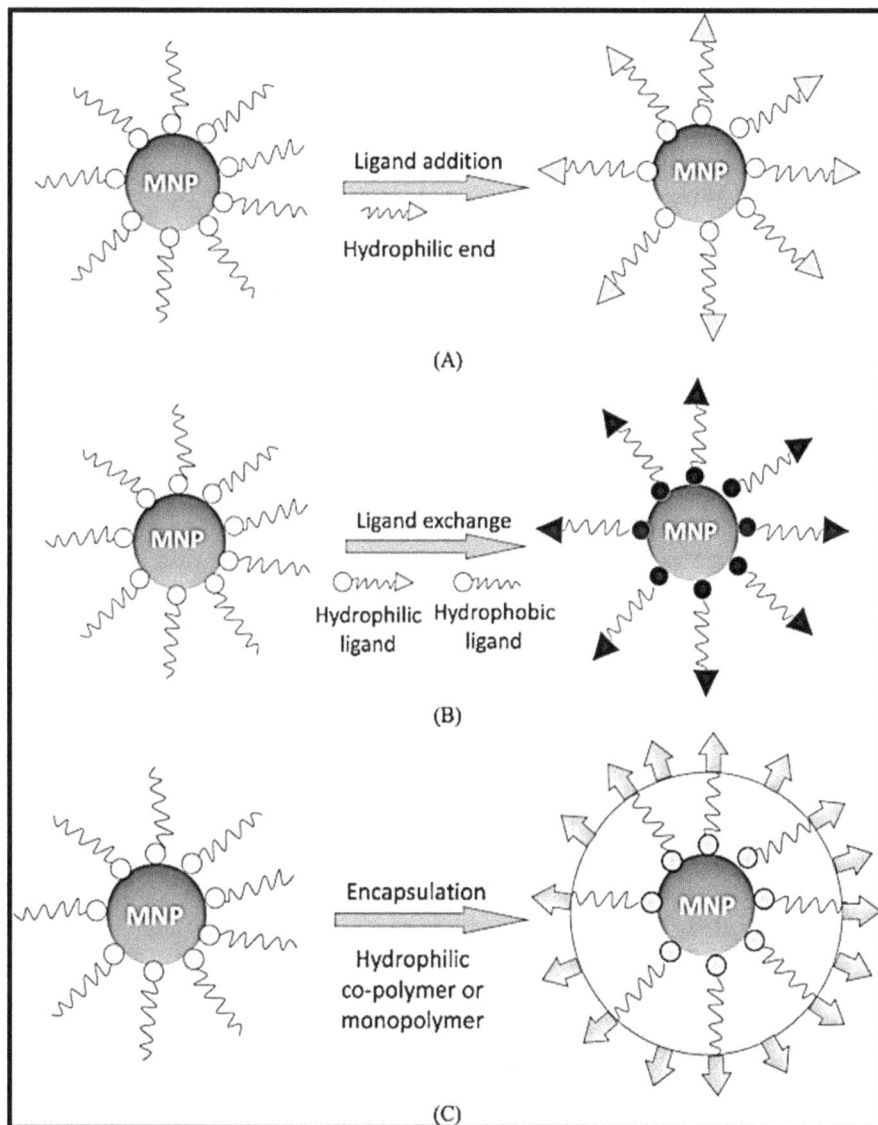

Figure 3.3. The mechanism of MNP surface FUCTN via (a) ligand addition, (b) ligand exchange, and (c) encapsulation. Reprinted from [4], Copyright (2023), with permission from Elsevier.

3.4 Magnetic properties of magnetic nanoparticles

The penetration of MFs into human tissues and the capacity to remotely sense or operate magnetic materials (MMs) for medical applications have been studied for centuries. A more current and significant use of these properties is in MRI, which is a NI imaging modality capable of offering HR functional images. Yet, the existing medical imaging potential can be significantly extended through the use of MNPs, which enhance the distinction between diseased and healthy tissues. To fully comprehend the benefits of MNPs as MRI CATs, we must have a fundamental knowledge of the concept of magnetism and MNP properties [14].

The classification of materials' magnetic properties is based on their magnetic susceptibility, χ, which is the ratio of the induced magnetism (M) to the applied magnetic field strength (H). For diamagnetic materials, the MM is antiparallel to H, resulting in very minimal and negative (−ve) values of χ ($1 \times 10^{-6} - 1 \times 10^{-3}$). For these types of MMs, when the external MF (EMF) is removed, their MP is not retained. Paramagnetic materials are defined as materials in which the MM is aligned parallel to H and the χ value is in the range of $10^{-6} - 10^{-1}$. In ferri- and ferromagnetic materials, there are coupling interactions between the material's electrons when their MMs are aligned parallel to H, which results in well-organized magnetic states like magnetic domains (MDs) and significant spontaneous magnetism. Materials' susceptibilities depend on the EMF, their atomic structure, and temperature [14].

Trivial sizes of ferri- and ferromagnetic materials (MNPs) in the range of tens of nanometers (nm) develop a single MM and hence sustain their unique huge magnetism. Yet, at appropriately elevated temperatures like the blocking temperature (T_B), there is an induction of free particle rotation, resulting in a loss of net magnetism in the absence of an EMF due to thermal energy [27, 28]. Depending on their natural magnetism, MNPs are grouped into paramagnetic (PM) and ferromagnetic types. These magnetic features are strongly impacted by the NP size, and when the size is less than a certain critical value, they form a single MD. These domains are areas where magnetons in a solitary volume of ferromagnetic materials align in a similar direction under the effect of an exchange force. SPM appears when the thickness of a single-magnetic-domain NP (between 3 and 50 nm, depending on the material) is also less than the critical value, and the coercive force is zero [29].

MNPs are the most commonly used superparamagnetic NPs (SNPs), and SPM is generally triggered by thermal influence, as depicted in figure 3.4(a). When these SNPs are dissolved in a fluid, they display nonmagnetic behavior in the absence of an EMF. Upon the application of a weak MF, thermal agitation can moderately impair the change of dipole moment (DP) orientation toward the MF. With increasing MF intensity, the SNPs progressively align. When it exceeds a defined value, the magnetism attains a saturated state where the SNPs are all aligned (see figure 3.4(b)). The improved Langevin function is used to define the superparamagnetic magnetism, which depends on the applied MF strength (H) and the M_s of the SNPs, as given by equation (3.1) [29]:

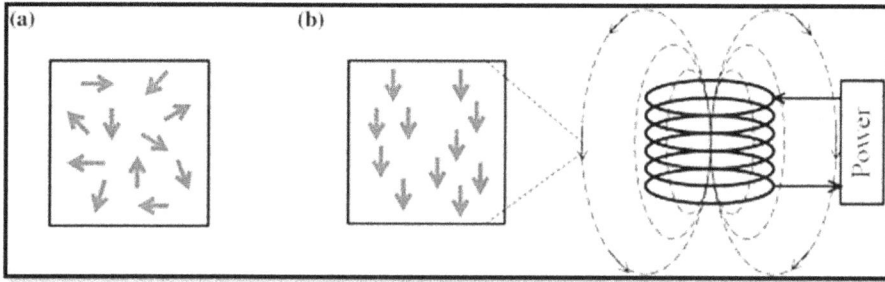

Figure 3.4. (a) NPs isolated in a fluid with nonmagnetic behavior in the absence of an EMF; (b) complete alignment of the NPs with the strong application of a sufficient MF, resulting in a DP which is aligned with the field, overcoming thermal agitation. Reprinted from [29], Copyright (2020), with permission from Elsevier.

$$M = M_s L\left(\frac{\mu_p H}{k_B T}\right) + \chi_a H. \tag{3.1}$$

Here, $L(x) = \cot - 1/x$, x, μ_p, and χ_a are the Langevin function, the Langevin factor, the average MM of individual SNPs, and the linear factor of the magnetic susceptibility, respectively. With these magnetic features, these NPs can produce an induced field when they are exposed to an EMF, which helps to reveal their location and allows them to be sensed remotely [29, 30].

In a study by Mohapatra *et al* [3], it was observed that MNPs had a central shell structure that was magnetically well-ordered, while the shell appeared to have a disordered structure identified as a spin canting/spin disorder film (figure 3.5(a)). It can be noted in figure 3.5(b) that the sizes of the various magnetic ferrites depended on the saturation magnetization (M_s) values obtained. The influence of surface spin canting reduced with increasing MNP size, while the M_s value improved rapidly to the substance's M value. This proportional association between the size of the particles and M_s may result in an improvement in the MRI signal intensity with increasing MNP size. There was a gradual increase in the R_2 values as the MNP size (MnFe$_2$O$_4$) increased from 6 to 12 nm (figure 3.5(c)). The MnFe$_2$O$_4$ NPs were seen to have a better effect on magnetic resonance (MR) contrast compared to Fe$_3$O$_4$, owing to their elevated M values. The CT M values can be controlled further by the introduction of various TMs (Ni, Co, Fe, and Mn) with diverse MMs into the host NPs (Fe$_3$O$_4$). The relationship between the influence of the M_s values and the R_2 RELVTY of the MNPs in a sequence of various magnetic ferrites and their multicore structures is shown in figure 3.5(d). It can also be seen in this figure that for an array of MNPs in the typical size range of 6–16 nm, there was a strong association between the R_2 RELVTY and the M_s values. The modulation of the M_s values substantially enhanced the R_2 values from less than 100 to nearly 700 mM^{-1} s^{-1} [3].

Particles maintain their colloidal stability and avoid aggregation, making them suitable for biomedical use, which is enabled by the superparamagnetic property marked by the lack of remnant magnetism following the removal of the EMF. Also,

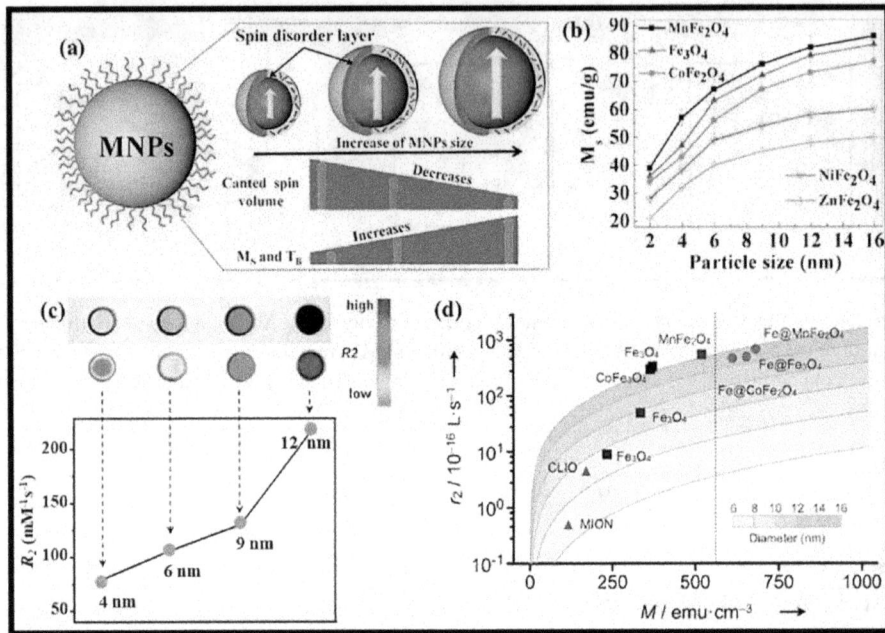

Figure 3.5. The influence of MNP size on T_2 MR contrast properties based on (a) a graphical representation of the spin canting influence on the M_s variation and T_B with varying MNP sizes, (b) the dependence of the MNP size on M_s, (c) the dependence of R_2 relaxivity value on $MnFe_2O_4$ NP size, and (d) the relationship between the R_2 relaxivity value and the MNP size. Reprinted from [3], Copyright (2023), with permission from Elsevier.

there are more developed magnetic susceptibilities resulting in coupling interactions within single MDs than in paramagnetic materials. With a further decrease in the particle size, there is an increase in the surface-to-volume ratio, resulting in surface effects such as spin canting, spin glass, and non-collinear spin behavior, which can considerably affect the material's MPs [27]. Given the MD concept, the critical size of a single domain is impacted by M_s, the crystal anisotropy strength, exchange force, the surface or domain wall energy, and the particle shape [30].

3.5 MRI mechanism

The diversity of the unique physicochemical properties of NP materials is impacted by size reduction down to the nanoscale. This change in properties affects the NPs' behavior in the biological microenvironment and enables NPs' interactions at the cellular and molecular levels. This is significant in that, to operate these imaging modalities at their optimum potential, the nanoprobes need tailored properties and should be engineered accordingly. The synthesis mode, surface chemistry, size, shape, magnetism, and anisotropy parameters can be used to control and boost nanoprobe quality [3].

The first clinical MRI was developed in 1980 at Nottingham and Aberdeen. It is a widely accessible and powerful medical tool [31]. Since the early 1980s, when MRIs were first approved for use in human patients, it has been based on the MM relaxation of water protons and their impact due to the application of an external MF. This relaxation is additionally affected by the application of MNPs in the direct microenvironment of the target tissue/organ. The NPs are dispersed in 3D and used as a CAT, as they create a resident MF gradient. This impacts the spin–lattice and spin–spin relaxation of protons in tissues/organs and permits brighter or darker signal contrast [3].

Atomic nuclei are comprised of neutrons and protons with a net positive charge. Atomic nuclei such as phosphorus (^{31}P) and hydrogen (^{1}H) nuclei have spins that depend on the number of protons. Based on a mathematical analogy, these nuclei are considered to spin about their own axes [31]. In the classical sense, spin describes the MM that emanates from or is related to a current loop produced by the spinning charged particles when a charge is present on the exterior of the particle surface, and this can be computed using equation (3.2) [32].

$$I = qV/2\pi r \qquad (3.2)$$

where q, r, and V are the particle charge, particle radius, and divergent velocity of a point on the particle surface, respectively [32].

For clinical MRI, the spin of interest is linked to the water proton. The MM is a vector quantity that is a product of the particle area and the current. This vector quantity has a direction that is linked to the particular momentum of the spinning particle, as represented by equation (3.3). It is well known that spins tend to line up with an EMF, in the same way that iron filings line up with an MF in free space. Relative to the direction of the applied field, there are two ideal spin configurations, namely spin up and spin down:

$$\mu = q/2m \times J \qquad (3.3)$$

where μ, J, q and m represent a DP μ with angular momentum (J), the charge, and the particle mass, respectively [32, 33].

MRI offers deep soft tissue imaging, permitting it to detect variations in structural and molecular abnormalities. MRI directly senses the presence of NPs via the effect of the MNPs' MF on the ^{1}H protons in the area [5]. MRI has polarization, encoding, and detection stages. In the first phase, the majority of the object's nuclear spins are polarized in an EMF (B_o) pointing in a certain direction. Constant radio-frequency (RF) pulses are used in assessing the MF (B_m) to create resonance absorption in protons, during which the MF gradient ($G_{x,y,z}$) is purposefully used to provide 3D encoding. Finally, via relaxation, the nuclei are returned to equilibrium with an energy loss via the discharge of their own RF signal, known as the free induction decay (FID) response signal. This is measured to recreate a 3D MR image of the object [34]. The net magnetism is determined by the MF strength (B_o), the gyromagnetic ratio (γ), the nuclear spin (I), and the inverse of the temperature [35].

The MM of water protons lines up parallel (lower energy state) and antiparallel (higher energy state) to the field B_o, with a unique parallel spin population of 0.001% in a static MF. This unique proton parallel to B_o recoils, precessing around B_o with an exclusive MM of M_z and a precession frequency or Larmor frequency of ω_o. This is given by equation (3.4) ($\gamma = 2.67 \times 10^8$ rad·s^{-1} T^{-1} for ^{1}H) [3, 36].

$$\omega_{\circ} = \gamma B_{\circ} \tag{3.4}$$

Moreover, the resonant RF pulse (perpendicular to $B_{\circ}$) irradiation leads to resonant excitation, which results in an improvement in the crosswise magnetism. The crosswise magnetism (M_{xy}) improves and undergoes precession in the crosswise plane. When the RF pulse is inhibited, the M_{xy} is progressively returned to its preliminary state of equilibrium by realigning to $B_{\circ}$. The spin–relaxation process involves M_z recovery, which arises when energy is lost from the excited state to its surroundings (spin–lattice relaxation), and M_{xy} decay, which arises when the precessing proton spins on the xy-plane, losing phase coherence due to the spin–spin interaction, as depicted by figure 3.6 (as adapted from Mohapatra *et al* [3]). These processes are defined by equations (3.5)– (3.9) [3, 36]:

$$M_{z(t)} = M_{z(0)}\left[1 - \exp\left(-\frac{t}{T_1}\right)\right] \quad \text{(longitudinal)} \tag{3.5}$$

$$R_1 = \frac{1}{T_1} = \frac{1}{T_1, \text{H}_2\text{O}} + r_1[C] \tag{3.6}$$

$$M_{xy(t)} = M_{xy(0)}\exp\left(-\frac{t}{T_2}\right) \quad \text{(transverse)} \tag{3.7}$$

$$R_2 = \frac{1}{T_2} = \frac{1}{T_2, \text{ H}_2\text{O}} + r_2[C] \tag{3.8}$$

$$\frac{1}{T_2^*} = \frac{1}{T_2} + \gamma \Delta B_{\circ}. \tag{3.9}$$

When the RF pulses are produced, they cause a significant excitation of proton nuclear rotation aligned antiparallel to $B_{\circ}$, thereby producing transverse magnetism while reducing longitudinal magnetism. Upon the cessation of the RF pulse, the

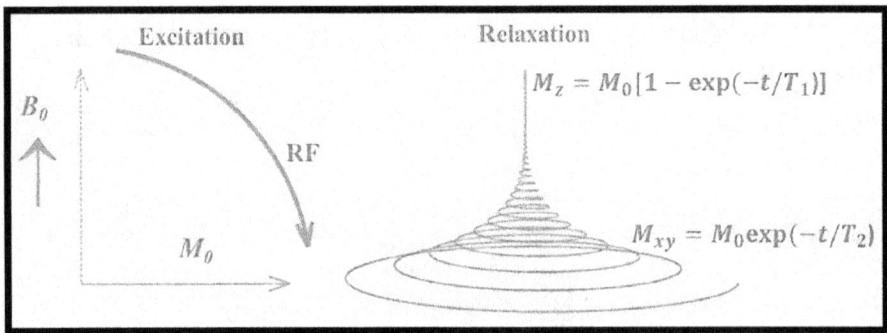

Figure 3.6. The longitudinal relaxation (LR) and transverse relaxivity (TR) of the net proton M. Reprinted from [3], Copyright (2023), with permission from Elsevier.

excited protons relax to the ground state and align parallel with B_o, thereby releasing the energy obtained from the RF pulse. This relaxation proceeds through the T_1 longitudinal relaxation (LR) (spin–lattice relaxation resulting in longitudinal magnetism recovery) and T_2 transverse relaxation (which causes transverse magnetism decay due to phase coherence loss between the proton nuclear spins). The recaptured M_z is defined by equation (3.5), where the T_1 RT is described by the time involved in recovering 63% of the unique longitudinal magnetism. For T_2 relaxation (spin–spin relaxation—equation (3.7)), in-phase protons are dephased with the decay of the transverse magnetism (M_{xy}). This decay procedure follows equation 3.6, where T_2 is the time required to decay to 37% of the unique M_{xy} value. Relaxation processes are records of magnetic resonance and are reconstructed into grayscale images. These MRI images are largely divided into imaging modes known as T_1 one-sided images and T_2 one-sided images (equation (3.8)). For T_1 one-sided MRI images, a quicker T_1 relaxation rate (RR) ($R_1 = r_1 = 1/T_1$ (S^{-1})) gives brighter contrast. For T_2 one-sided MRI images, a swifter T_2 RR ($R_2 = r_2 = 1/T_2$ (S^{-1})) yields darker contrast (equation (3.5)). Generally, spins degenerate quicker than T_2 owing to the MF inconsistency created by T_2 CATs. The actual RT (T_2^*) is given by equation 3.9, where ΔB is the difference in the local MF strength due to the inconsistency [2, 7, 36–39].

The images are obtained using the arrangement shown in figure 3.7(a), which shows a patient or biomedical sample placed in the interior of the MR scanner. The MRI arrangement consists of a core magnet, gradient coils, RF coils, and a

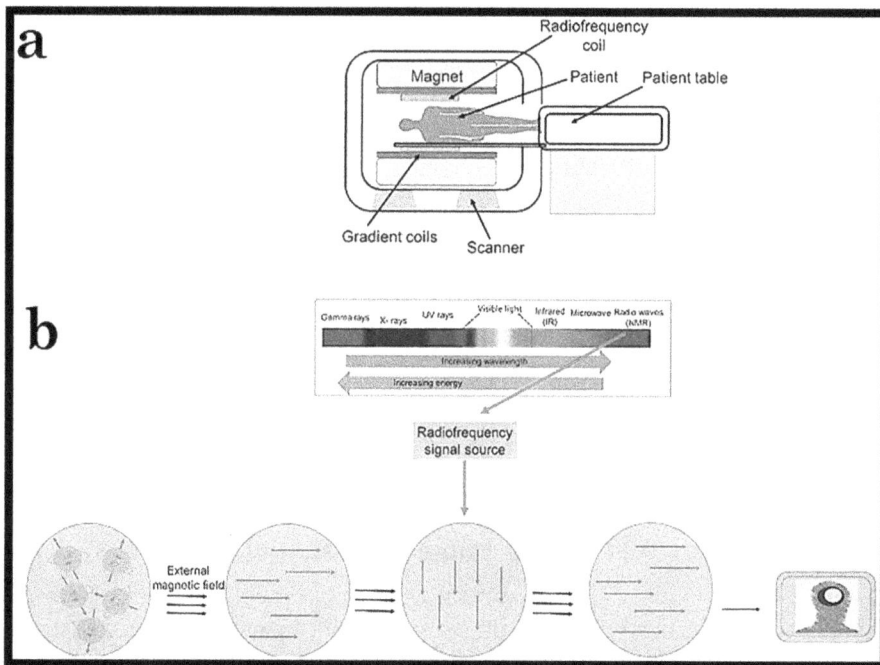

Figure 3.7. (a) Graphical illustration of an MRI scanner and (b) MRI signal acquisition. Reproduced from [15]. CC BY 4.0.

computer system that operates and interfaces different apparatuses. The biomedical sample or patient placed in the interior of the scanner is exposed to an EMF, generally between 0.5 and 3 T (in clinical settings) and up to 10.5 T in research studies [15, 40]. This lines up the water protons' magnetic poles with the applied MF, and the images are produced by regulating proton relaxation in the presence of an EMF. The hydrogen nuclear spin orientations in the presence of a strong MF are parallel and antiparallel to the MF direction [15, 40]. The number of protons excited in the high-energy state is amplified by the resonant RF radiation, which transfers energy to the protons. Upon the cessation of the RF pulse, the MM (spin) of the protons returns to its unique relaxation condition. Two vital variables that impact signal intensity and image contrast are the repetition time (TR, or the amount of time between RF pulse repetitions) and the echo time (TE or the time between the RF pulse and the echo signal peak). TE (>10 ms) and TR (>250 ms) are carefully selected short times for T_1-weighted images (WIs), while long TR (>2000 ms) and long TE (>60 ms) are utilized to produce images using T_2 conditions (figure 3.7(b)) [15, 40].

In medical examinations, positive contrast is generally preferred to negative contrast because it circumvents the potential misperception of the signal decay produced by the negative CATs with signal cavities caused by MF inhomogeneities arising from metal prostheses, air, etc [13]. T_1 and T_2 CATs generally used are paramagnetic complexes such as gadolinium (Gd) diethylenetriamine penta-acetic acid (Gd-DTPA) and MNPs. Strong contrast can be obtained by using metal oxides (MeO) in T_2 one-sided images, and these MNPs are taken up non-specifically by the RES, consisting of the lymph nodes (LNs), spleen, and liver, diminishing the effectiveness of their recognition. Yet, effective T_2 CATs include manganese ferrite or graphene oxide nanohybrids created using the thermal decomposition method. They show no substantial cytotoxicity and hemolytic activity in MRI experiments *InVi* and *in vitro*. Various types of molecules, such as carcinoembryonic antigen (CEA), epidermal growth factor (EGF), and Herceptin, can be used to modify MNPs for the efficient recognition of tumors [4]. The key role of T_1 and T_2 CATs in MRI is to reduce the selectivity of the RTs of the water protons in the area of interest, hence providing improved contrast for body areas. Contrast is improved when tissue has a developed affinity for the CATs or increased vascularity. Unhealthy tissues like cancers are metabolically diverse from healthy tissues and take up these CATs in different ways, thereby resulting in contrast in MRI images. T_1 WIs show structures well and are chosen when a distinct image of a structure is needed [7].

3.6 Recent MRI studies using magnetic nanoparticles

MRI stands out among medical analytical approaches owing to its exclusive blend of possibilities, such as its NI application, absence of ionizing radiation, outstanding image quality, and capacity to deliver structural and practical information. The MRI signal generally emanates from water molecule protons, while the image contrast is created by the variance in intensity between the various tissues, which can depend on the RTs (T_1 and T_2), the mobility of water molecules in the interiors of individual tissues, and the concentration. Furthermore, the image contrast can be further improved using CATs. The capability of CATs to reduce the RTs (T_1 or T_2) is given by the factors known as the RELVTY (r_1 or r_2), expressed in $mM^{-1} s^{-1}$ [13, 41].

To develop access to MRI for patient neuroimaging and diagnosis, low-field MRI, which is a type of MRI scanner that functions at a decreased MF of 1–100 mT, is applied. Compared to characteristic medical MRI scanners functioning at 1.5–3 T, this low MF can be produced using permanent magnets and hence requires a reduced amount of space, power, and associated infrastructure compared to medical scanners that employ cryogenic superconducting magnets. Also, the signal-to-noise ratio of MRI scales roughly quadratically with the field strength. MRI scanners do not precisely sense MNPs but they enhance contrast in the reconstructed MR images as the signal intensifies (positive CATs) or as the signal decreases (negative CATs). Currently, hardware improvements and cutting-edge image modernization, such as deep learning and postprocessing, have led to the creation of novel low-field scanners. Despite their signal-to-noise ratio, hardware and software advancements allow low-field scanners to deliver diagnostically useful data within realistic scan times, notwithstanding their lower spatial resolution when compared to the resolution offered at clinical field strengths [42, 43].

The degree of MRI improvement brought about by MNPs is reliant on their size, structure, aggregation degree, and peripheral effects. Their size is a key factor in improving the TR T_2, while surface contact affects the LR T_1 [43]. MRI contrast can be greatly impacted by the MNP size, and according to theoretical research about the impact of NP size on T_2 RELVTY, the three size regimes are the static dephasing regime (SDR), the echo-limiting regime (ELR), and the motional average regime (MAR). As the NP size increases, R_2 improves in the MAR and attains its maximum plateau (SDR). A decrease in the R_2 (ELR) is reported with a further increase in NP size. Though NPs in the SDR have the maximum R_2, the size of the MNPs employed in MRI applications is generally within the MAR because the NPs in the SDR suffer from random accretion enabled by strong ferromagnetic dipolar interactions. The MNPs' composition is another critical factor for MRI contrast impacts. The magnetism of NPs can readily be altered by the introduction of transition-metal dopants. The impact of metal doping with metal ferrite NPs involves the replacement of the Fe^{2+} in the octahedral sites of 12 nm NPs with TMs dopants such as Ni^{2+}, Mn^{2+}, or Co^{2+}. In addition, the doping approach is used to improve T_1 contrast impacts by doping rare-earth metals into ferrite NPs. Gd-implanted FeO (Gd–FeO) NPs have been used to disrupt the spin ordering of the FeO NPs. The NPs' spin orientation and surface spins are asymmetrical; however, the internal spins, which are magnetically disordered, are affected by the Gd dopants, and this results in low r_2/r_1 values. T_2 RELVTY is impacted by the NP's M_s value and radius (r). Hence, for an ambitious M_s value, a highly developed T_2 RELVTY can be achieved by altering the radius of the MNPs, which is mainly shape-dependent [2].

The contrast is improved by enhancing the photon RT with different metallic NPs like manganese (Mn), Gd, and FeO NPs (such as Fe_2O_3 and Fe_3O_4 NPs) applied as CATs. Fe_2O_3, Fe_3O_4 NPs, and Gd are chelated organic complexes, which have been significantly utilized as CATs in MRI due to their capacity to isolate oxygen (O) and Fe in the human body. They can be detached securely and used for the metabolic and methodical transport of O [44]. Since MNPs possess low cytotoxicity, they have been accepted by the United States Food and Drug Administration (USFDA) for medical uses. Fe_3O_4 NPs, Gd, and Mn have high molar T_2 relaxivities. Moreover, Mn and Gd-based MRI CATs have been researched for *InVi* and *in vitro* imaging uses and accepted by the USFDA for human applications. These are paramagnetic metal ions (MIs) with seven unpaired electrons in the outer shells of lanthanides and TMs, which combine with a long electron spin RT to make these MIs effective relaxation-improving CATs. The five MRI CATs accepted by the USFDA are based on Gd^{3+} ions because these materials have a strong ability to catalyze water-signal relaxation to create a positive contrast in MRI; in addition, they have strong PM and are inert and stable in the body. The two major groups of Gd^{3+} chelates applied in medicine are the cyclic (macrocyclic ligand) and acyclic (acyclic ligand) groups. The drawbacks of traditional Gd chelates include the absence of analytical specificity and the toxicity associated with their application due to the unanticipated release of free Gd ions. MNPs have emerged as a capable substitute that can overcome these drawbacks. The high 3D resolution provided by MRI and the improved negative contrast offered by MNPs when utilized with T_2-weighted pulse sequences make MRI an outstanding imaging modality for the detection of disease tumors as small as 2–3 mm in clinical applications [13, 43, 45–47].

Cancer is a key cause of death globally. MRI has been applied in liver, colon, breast, and lung cancer screening programs. LN staging is indispensable for recognizing cancer forms and selecting optimal treatment strategies. Understanding how well a patient will survive is supported by knowing how many LNs are affected by the metastatic process and how various LNs are impacted. LN staging is frequently supported by protected LN imaging or by medical segmentation, followed by a histological investigation of limited neighboring LNs that strengthen the unique tumor site. Low recognition rates for extremely small MeTas and erroneous positive evaluations of swollen LNs as MeTas are typical challenges in normal NI LN staging by size. Moreover, the surgical dissection procedure is invasive and generally hard to apply. Surgery only affects certain areas neighboring the main tumor, which may cause it to miss distant LN MeTas that often occur in the pararectal and common iliac areas and thus underestimate the presence of remote LN MeTas [15].

In a study by Oberdick *et al*, SFeO NP CATs such as USFDA-approved ferumoxytol (FEYL) were investigated as positive T_1 contrast agents in a low-field MRI at 64 mT. Based on a transmission electron microscopy (TEM) analysis (figures 3.8(a) and (b)), the particles used as CATs in this study (figure 3.8(a)) were found to be spherical monodisperse SFeO NPs which were stabilized by coating them with carboxylic acid. Their average particle size was about 5–16 nm. The isolated core made of SFeO NPs was well separated owing to the blend of electrostatic and steric forces in a water-soluble medium. Figure 3.8(b) shows USFDA-approved FEYL, which is applied for the treatment of iron deficiency disorders. It comprises an SFeO NP core enclosed by a carbohydrate covering. Figures 3.8(c) and (d) display the ambient temperature (about 22 °C) assessment of

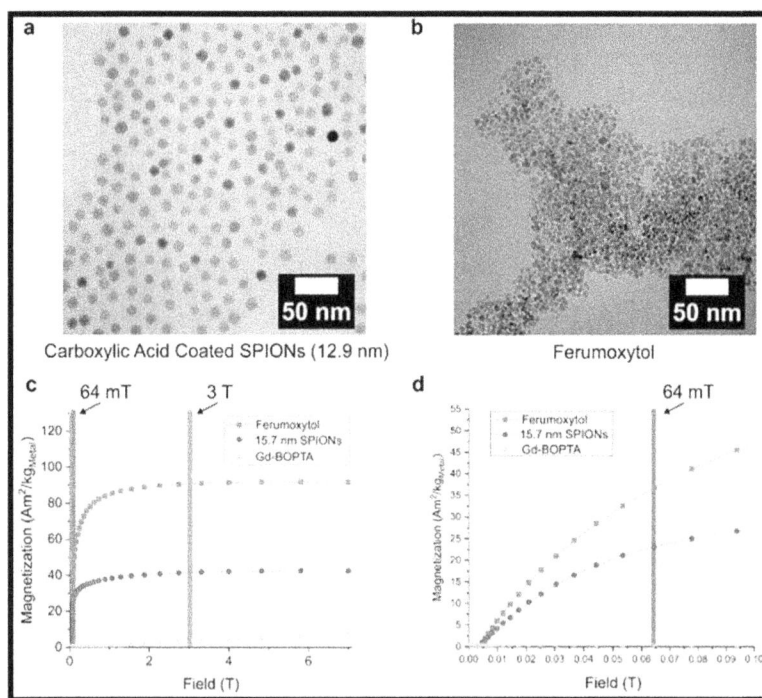

Figure 3.8. TEM images of (a) 13 nm SFeO NPs covered with carboxylic acid, (b) FEYL. (c) Plot of M_s versus applied MF for FEYL, 16 nm SFeO NPs, and Gd–BOPTA with vertical bars displaying field areas corresponding to 64 mT (low-field MRI) and 3 T (typical medical MF). (d) Extended area of M_s versus applied MF in the low-field regime. Reproduced from [42]. CC BY 4.0.

the M_s as a function of the applied MF for the SFeO NPs CATs, 16 nm carboxylic-acid-covered NPs, and FEYL. Based on figure 3.8(d), the M_s value obtained for the bulk FeO NPs was 92 Am^2 kg^{-1}. A reduced M_s value ranging from about 40–75 Am^2 kg^{-1} was obtained for the carboxylic-acid-covered particles. Gadobenate dimeglumine (Gd–BOPTA) showed paramagnetic behavior and a direct linear improvement in M_s as a function of the applied MF. The MRI information at 3 T and 64 mT for the CATs (16 nm carboxylic-acid-covered NPs, FEYL, and Gd–BOPTA) is displayed in figure 3.9, as adapted from Oberdick *et al* [42]. The LR curves employ signal intensity taken from a 64 mT inversion recovery fast spin echo (FSE) sequence. The 16 nm SFeO NPs were the most effective at decreasing T_1, followed by ferumoxytol at 64 mT. The slowest degree of T_1 relaxation was observed in Gd–BOPTA, as shown in figure 3.9(a). There was a substantial reduction in the T_1 relaxation of the Fe$_3$O$_4$ NP CATs compared with the T_1 relaxation of the Gd–BOPTA CATs at 3 T. Based on figure 3.9(c), FEYL and 16 nm NPs had signal intensities that were about five and six times greater than that of Gd–BOPTA because they relaxed longitudinal magnetism much more effectively. Therefore, they appeared brighter in comparison to Gd–BOPTA. Figure 3.9(d) displays an MRI taken at 3 T alongside the plots of the standardized intensity images. There was a

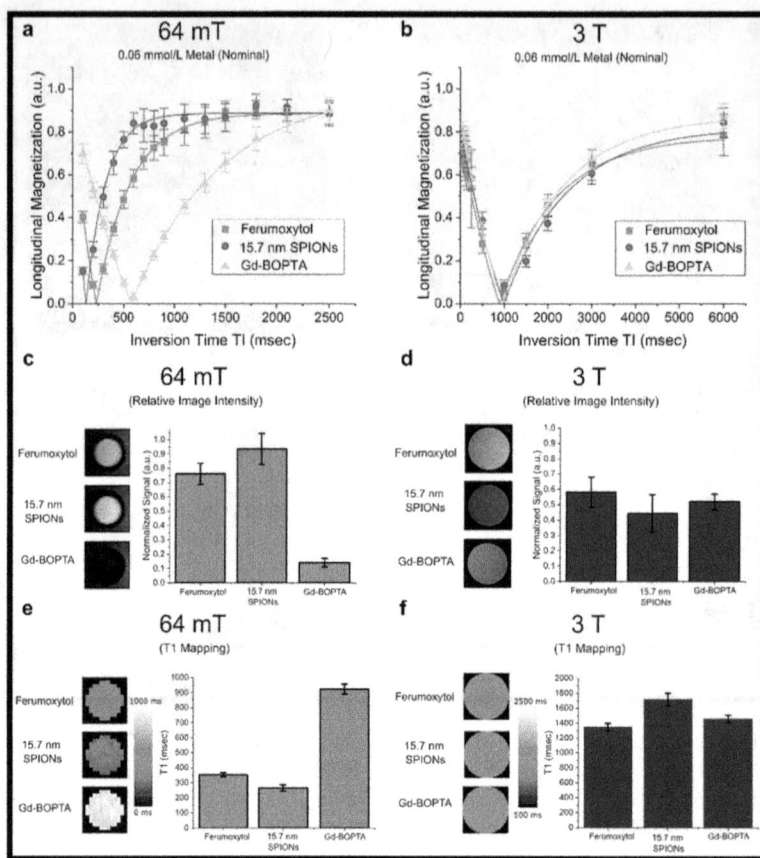

Figure 3.9. Inversion recovery curves for 16 nm Fe_3O_4 NPs, Gd–BOPTA, and FEYL at (a) 64 mT and (b) 3 T. Normalized image intensities corresponding to (c) inversion times (TIs) of 700 ms obtained using 64 mT MRI and (d) 2000 ms obtained using 3 T MRI. T1 maps corresponding to the areas of interest (AOIs) at (e) 64 mT and (f) 3 T. Reproduced from [42]. CC BY 4.0.

slight distinction in contrast as a function of the additional agent. Figures 3.9(e) and (f) show a quantitative comparison of T_1 maps for the CATs at 64 mT and 3 T. At 64 mT, the nominal metal concentration of 0.06 mmol L^{-1}, FEYL, and 16 nm Fe_3O_4 NPs had T_1 values of 353 and 266 ms, respectively, which were 38% and 29% of the Gd–BOPTA value at 64 mT. There was less differentiation of the T_1 values between the CATs at 3 T, and the values obtained were 1347, 1716, and 1460 ms for FEYL, 16 nm Fe_3O_4 NPs, and Gd–BOPTA, respectively. An improved T_1 RT led to enhanced contrast and visibility of the tissues in the MRI scans using SFeO NPs [42].

In a study by Yallapu *et al* (2010), PEG and oleic acid were functionalized on Fe_3O_4 NPs to create OE-20, OE-40, and OE-80 MNPs (PEG-OA-MNPs), which were explored as MRI agents for *InVi* application. The MR signal intensities were substantially reduced as the particle concentrations (μg Fe ml^{-1}) in phantom gels were increased, especially in OE-80 MNPs at 20 and 50 μg Fe ml^{-1} gel, as displayed

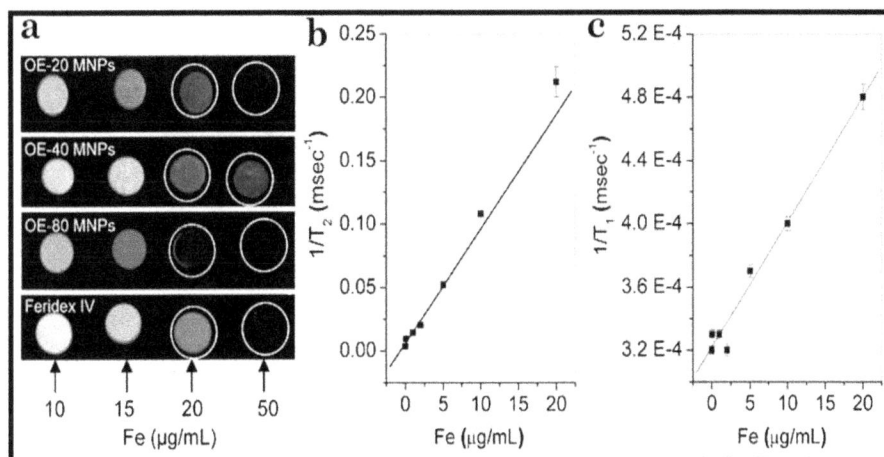

Figure 3.10. MRI of (a) contrast at changing Fe concentrations for PEG-OA-MNPs and Feridex (VI) in agar gels. (b) T_2 RELVTY and (c) T_1 RELVTY for OE-80-MNPs. Reprinted by permission from Springer Nature Customer Service Centre GmbH: [Springer Nature Link] Pharmaceutical Research] [48], Copyright (2010).

in figure 3.10(a). Based on figures 3.10(b) and (c), the RELVTY rates r_1 (T_1 RELVTY) and r_2 (T_2 RELVTY) were noted to vary widely for the various formulations assessed using $1/T_1$ and $1/T_2$ [48].

Employing hydrothermal and sonochemical methods, amino phosphate (diethylenetriamine penta(methylene phosphonic acid), DTPMP) was coated on Fe_3O_4 NPs and used as a CAT in an MRI study. The least detectable cells and the association between the number of cells inoculated and the signal obtained from a phantom experiment are shown in figure 3.11, as adapted from Neto et al [49]. The signals at the individual inoculation sites were predicted by computing the signal in a rectangular area of interest (AOI) that included the inoculation site of 230 pixels. A strong linear correlation between the projected value and the signal was noted ($r = 0.87$, $P < 0.005$). Hence, inoculation of 5×10^5 cells was easily spotted, which suggested that the detection limit of the approach was lower. The outline of the bright area, with a signal originating from the right and left of the inoculation (along the axis perpendicular to the main field), was obtained when the MF was removed (figures 3.11(a)–(c)). Additionally, RELVTY and cytotoxicity analyses established the potential of MRI CT using the DTPMP-coated Fe_3O_4 NPs. This confirmed a non-cytotoxic profile and high TR values of 357–417 m M^{-1} s^{-1} [49].

A study by Hachani et al examined the interaction between 3,4-dihydroxy hydrocinnamic acid (DHCA) coated on Fe_3O_4 NPs that had the required features for T_2-weighted MRI and bone marrow produced by key human mesenchymal stem cells (hMSCs). TEM analysis showed (figure 3.12, as adapted from Hachani et al [50]) that the spherical Fe_3O_4 NPs had an average particle diameter of about 17 ± 2 nm. Based on cell uptake, as imaged by TEM at incubation times of 1–24 h and a concentration of 50 µg ml^{-1}, it was revealed that this concentration can be considered non-toxic and safe, with no impact on the morphology, viability, or mitochondrial

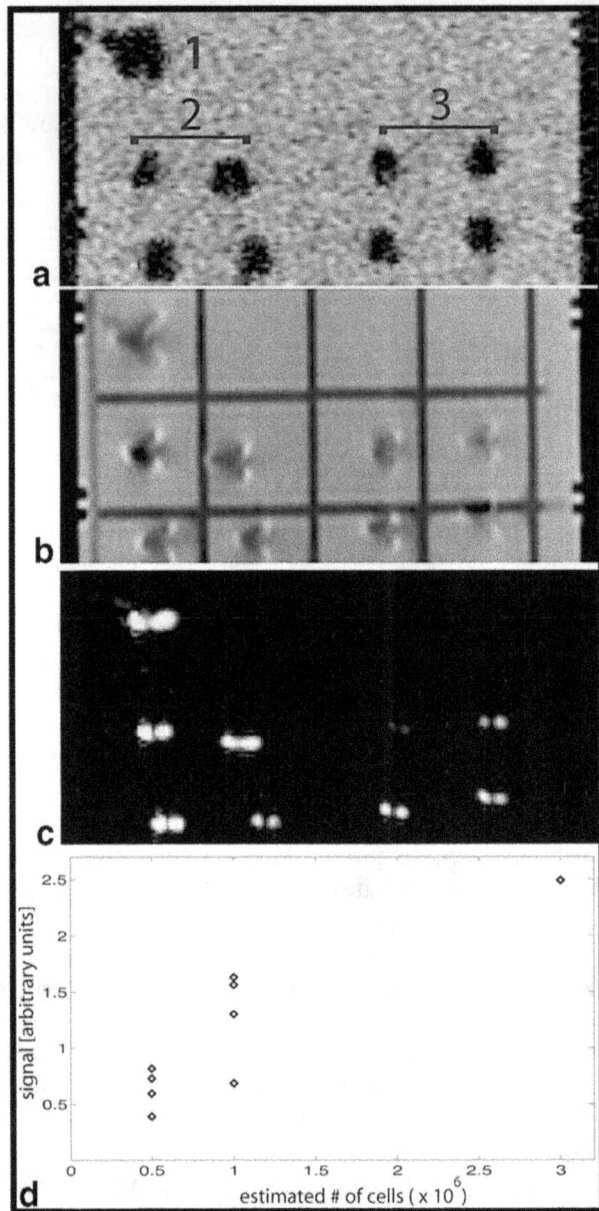

Figure 3.11. Assessment of off-resonance (OR) MRI of magnetically labeled cells and traditional gradient echo (GE) images (a) showing cavities at inoculation sites comprising (1) three million cells, (2) one million cells, and (3) and 0.5 million cells. (b) On-resonance projection showing cavities but with less contrast owing to background signal. (c) OR projection (−800 Hz) displaying outstanding contrast and (d) a graph of the signal shown in (c), calculated by computing the signal in a trivial AOI comprising individual bright spots versus the projected cell number. Reprinted from [49], Copyright (2021), with permission from Elsevier.

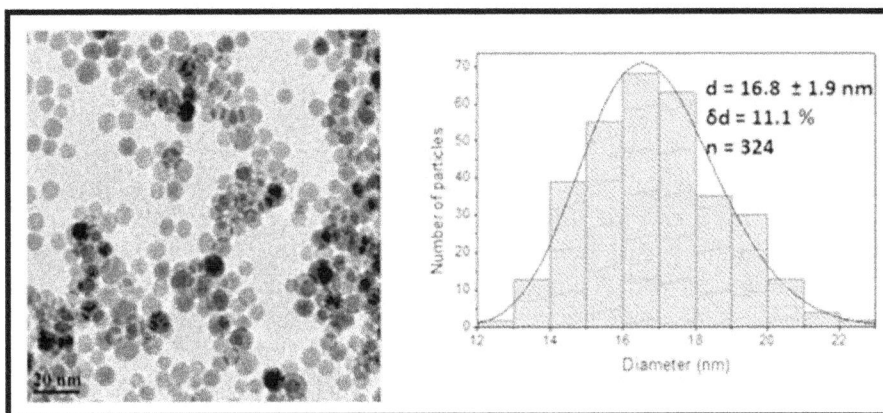

Figure 3.12. TEM image (left) and the size distribution of Fe_3O_4 NPs (right). Reproduced from [50]. CC BY 4.0.

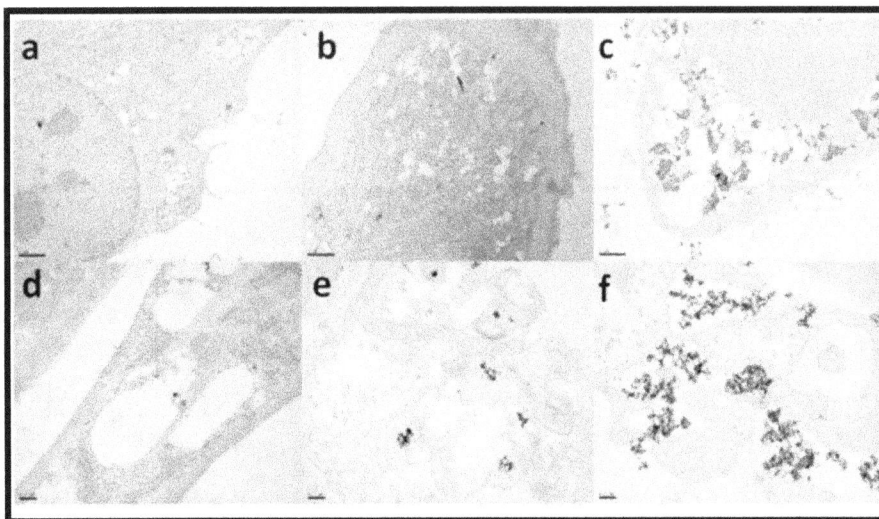

Figure 3.13. TEM images of hMSCs cultured with Fe_3O_4 NPs at 50 μg ml^{-1} for (a and d) I hr, (h and e) 4 h, and (c and f) 24 h. Reproduced from [50]. CC BY 4.0.

health of the cells and no associated reactive oxygen species production. Based on the images in figure 3.13, as adapted from Hachani *et al* [50], there was a positive uptake of the Fe_3O_4 NPs by hMSCs. This process was relatively slow after incubation times of 1 and 4 h, based on the smaller numbers of Fe_3O_4 NPs observed. After 24 h, substantial numbers of Fe_3O_4 NPs were visible inside the cells and on their surfaces. The RR constant R was linearly reliant on the Fe concentration. The NPs were observed to have r_1 and r_2 values of 0.78 and 142.2 mM^{-1} s^{-1}, respectively. The obtained r_2/r_1 values of 182 confirmed the potential for the Fe_3O_4–DHCA NPs to be applied as a T_2-weighted MRI CAT. The *InVi* performance of the Fe_3O_4–DHCA NP

CAT was assessed in the livers of six Swiss mice as models using T_2-weighted MRI. It was found that there was substantial signal attenuation in the liver area due to the Fe3O4–DHCA NPs, as displayed in figure 3.14. The *InVi* and *in vitro* studies showed that the Fe_3O_4 NPs were biocompatible and could create substantial contrast improvement in T_2-weighted MRI [50].

A study by Hossein *et al* [51] explored the use of MNPs prepared using the polyol approach and functionalized with folate to improve their performance as an MRI CAT specifically for breast cancer detection. It was found in this study that the MNPs functionalized with folate enhanced cellular uptake in the folate receptor-positive MCF-7 cancer cells, hence improving the imaging contrast. Figure 3.15 shows the association between the MR signal strength produced by the NPs and various sample concentrations, as adapted from Hossein *et al* [51]. Hence, enhancing the NP concentration within the MCF-7 cells substantially reduced the MRI signal linked with the T_2 RT. This led to negative contrast in the T_2 WIs. This study shows that the biocompatibility and efficiency of NPs improved the MRI signal, suggesting their possible use in improving the accuracy of breast cancer detection [51].

SFeO NPs functionalized with a lipid matrix were explored for use as an MRI CAT. A key finding from this study was the direct dependence of T_2^{-1} and T_1^{-1} on the Fe concentration, which was noted for all SFeO NP@lipid samples, thereby providing a secondary sign of the colloidal stability of the SFeO NP@lipid

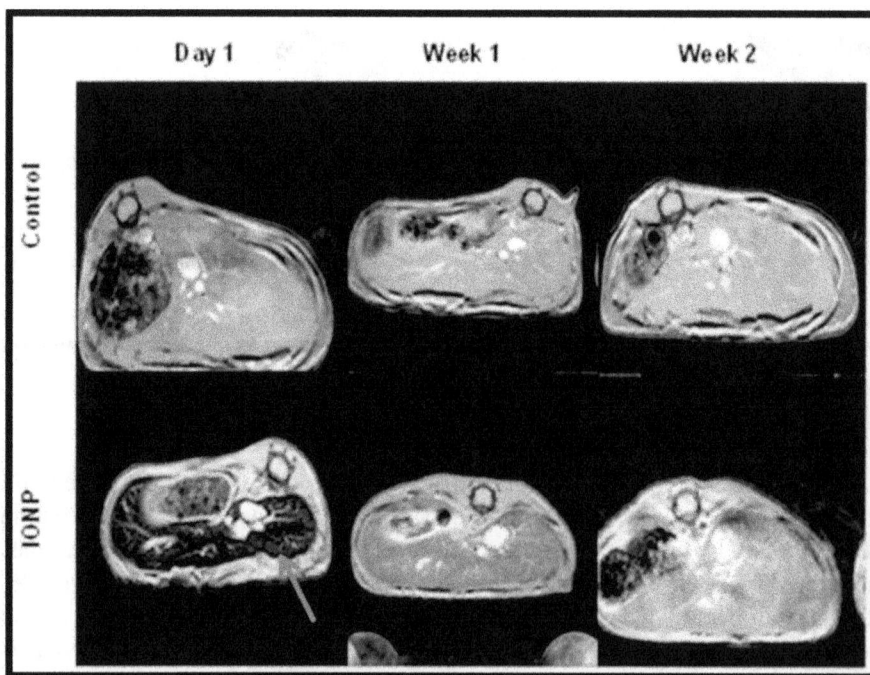

Figure 3.14. *InVi* T_2*-weighted MRI showing the central assessment of the mouse liver region after two weeks of injection of Fe_3O_4–DHCA NPs. Reproduced from [50]. CC BY 4.0.

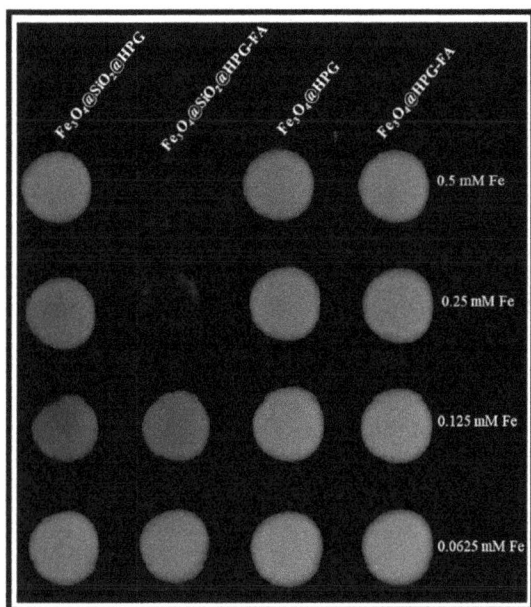

Figure 3.15. T_2 WIs of MCF-7 cells after treatment with various concentrations of CAT consisting of Fe_3O_4 doped with folate. Reproduced from [51]. CC BY 4.0.

dispersion. In addition, due to the superparamagnetic characteristics of SFeO NPs@lipid, they acted predominantly as a T_2 CAT. The r_1 value showed a strong reduction at a low magnetic loading of 0.1%–0.5%. Elevated levels of polyoxy-ethylene sorbitan monooleate (Tween80) had a negative impact on the r_2 RELVTY rate. As the percentage of Tween80 reduced, the r_2 values improved. Conclusively, the prepared SFeO NPs@lipid nanocomposite (NC) showed better performance in MRI application [52].

Employing the solvothermal approach, MNPs were prepared and used as an MRI CAT. *InVi* and *in vitro* MRI experiments showed that MF treatment resulted in homogeneous MNPs with sizes of about 5 nm, which substantially enhanced the T_1 contrast [53]. The MNPs were applied as a CAT to reduce the T_2 RT in the MR imaging of human prostate carcinoma DU-145 cells. In this study, no observed cytotoxicity was noted on two occasions at the target intracellular Fe concentration. There was a direct correlation between the reduced T_2 RT and the intracellular Fe concentration. This direct correlation is beneficial for the imaging and treatment of cancer cells [54].

3.7 Conclusions

MRI is a capable imaging approach owing to its ability to image structural and molecular status. It is a NI and nondestructive approach that can provide 3D images of living organisms. The application of MNPs in CATs enables substantial improvements in MRI specificity and sensitivity, since the resulting CATs modify the fundamental features of the tissues inside living organisms and thereby improve the image data. CATs enhance the signal and contrast of MRI by accelerating proton spin relaxation. MNP-based CATs provide improved MR signals and contrast when compared with those of mMRI CATs due to their improved MMs, extended blood circulation times, and extraordinary density of metal per NP.

References

[1] Rümenapp C, Gleich B and Haase A 2012 Magnetic nanoparticles in magnetic resonance imaging and diagnostics *Pharm. Res.* **29** 1165–79

[2] Shin T, Choi Y, Kim S and Cheon J 2015 Recent advances in magnetic nanoparticle-based multi-modal imaging *Chem. Soc. Rev.* **44** 4501–16

[3] Mohapatra J, Nigam S, George J, Arellano A, Wang P and Liu J 2023 Principles and applications of magnetic nanomaterials in magnetically guided bioimaging *Mater. Today Phys.* **32** 101003

[4] Gupta I, Sirohi S and Roy K 2023 Strategies for functionalization of magnetic nanoparticles for biomedical applications *Mater. Today Proc.* **72** 2757–67

[5] Rezaei B, Tay Z, Mostufa S, Manzari O, Azizi E, Ciannella S, Li C, Zeng M, Gomez-Pastora J and Wu K 2024 Magnetic nanoparticles for magnetic particle imaging (MPI): design and applications *Nanoscale* **16** 11802–24

[6] Materón E, Miyazaki C, Carr O, Joshi N, Picciani P, Dalmaschio C, Davis F and Shimizu F 2021 Magnetic nanoparticles in biomedical applications: a review *Appl. Surf. Sci. Adv.* **6** 100163

[7] Estelrich J, Sánchez-Martín M and Busquets M 2015 Nanoparticles in magnetic resonance imaging: from simple to dual contrast agents *Int. J. Nanomed.* **10** 1727–41

[8] Konnova S and Rozhina E 2024 Magnetic nanoparticles for biomedical and imaging applications *Int. J. Mol. Sci.* **25** 5847

[9] Aigbe U, Ukhurebor K, Onyancha R, Osibote O, Kusuma H and Darmokoesoemo H 2022 Measuring the velocity profile of spinning particles and its impact on Cr (VI) sequestration *Chem. Eng. Process.-Process Intensif.* **178** 109013

[10] Flores-Rojas G, López-Saucedo F, Vera-Graziano R, Mendizabal E and Bucio E 2022 Magnetic nanoparticles for medical applications: updated review *Macromol* **2** 374–90

[11] Zhang H, Liu X and Fan H 2023 Advances in magnetic nanoparticle-based magnetic resonance imaging contrast agents *Nano Res.* **16** 12531–42

[12] Popescu R, Andronescu E and Vasile B 2019 Recent advances in magnetite nanoparticle functionalization for nanomedicine *Nanomaterials* **9** 1791

[13] Avasthi A, Caro C, Pozo Torres E, Leal M and García Martín M 2020 Magnetic nanoparticles as MRI contrast agents *Top. Curr. Chem.* **378** 49–91

[14] Sun C, Lee J and Zhang M 2008 Magnetic nanoparticles in MR imaging and drug delivery *Adv. Drug Deliv. Rev.* **60** 1252–65

[15] Rahman M 2023 Magnetic resonance imaging and iron-oxide nanoparticles in the era of personalized medicine *Nanotheranostics* **7** 424

[16] Shukla S, Khan R and Daverey A 2021 Synthesis and characterization of magnetic nanoparticles, and their applications in wastewater treatment: a review *Environ. Technol. Innov.* **24** 101924

[17] Stiufiuc G and Stiufiuc R 2024 Magnetic nanoparticles: synthesis, characterization, and their use in biomedical field *Appl. Sci.* **14** 1623

[18] Sharma S, Sharma H and Sharma R 2024 A review on functionalization and potential application spectrum of magnetic nanoparticles (MNPs) based systems *Chem. Inorg. Mater.* **2** 100035

[19] Wu W, He Q and Jiang C 2008 Magnetic iron oxide nanoparticles: synthesis and surface functionalization strategies *Nanoscale Res. Lett.* **3** 397–415

[20] Lee J, Kim J and Cheon J 2013 Magnetic nanoparticles for multi-imaging and drug delivery *Mol. Cells* **35** 274–84

[21] Zhang K, Song X, Liu M, Chen M, Li J and Han J 2023 Review on the use of magnetic nanoparticles in the detection of environmental pollutants *Water* **15** 3077

[22] Comanescu C 2022 Magnetic nanoparticles: current advances in nanomedicine, drug delivery and MRI *Chemistry* **4** 872–930

[23] Yaashikaa P and Kumar P 2022 Fabrication and characterization of magnetic nanomaterials for the removal of toxic pollutants from water environment: a review *Chemosphere* **303** 135067

[24] Bohara R, Thorat N and Pawar S 2016 Role of functionalization: strategies to explore potential nano-bio applications of magnetic nanoparticles *RSC Adv.* **6** 43989–4012

[25] Zamay G, Zamay T, Lukyanenko K and Kichkailo A 2020 Aptamers increase biocompatibility and reduce the toxicity of magnetic nanoparticles used in biomedicine *Biomedicines* **8** 59

[26] Nguyen D and Kim K 2014 Functionalization of magnetic nanoparticles for biomedical applications *Korean J. Chem. Eng.* **31** 1289–305

[27] Lu A, Salabas E and Schüth F 2007 Magnetic nanoparticles: synthesis, protection, functionalization, and application *Angew. Chem. Int. Ed.* **46** 1222–44

[28] Biehl P, Von der Lühe M, Dutz S and Schacher F 2018 Synthesis, characterization, and applications of magnetic nanoparticles featuring polyzwitterionic coatings *Polymers* **10** 91

[29] Zhou K, Zhou X, Liu J and Huang Z 2020 Application of magnetic nanoparticles in petroleum industry: a review *J. Petrol. Sci. Eng.* **188** 106943

[30] Akbarzadeh A, Samiei M and Davaran S 2012 Magnetic nanoparticles: preparation, physical properties, and applications in biomedicine *Nanoscale Res. Lett.* **7** 1–13

[31] Grover V, Tognarelli J, Crossey M, Cox I, Taylor-Robinson S and McPhail M 2015 Magnetic resonance imaging: principles and techniques: lessons for clinicians *J. Clin. Exp. Hepatol.* **5** 246–55

[32] McGowan J 2008 Basic principles of magnetic resonance imaging *Neuroimaging Clin. N. Am.* **18** 623–36

[33] Onyancha R, Oyomo B, Aigbe U and Ukhurebor K 2023 Application of magnetic nanomaterials as drug and gene delivery agent *Magnetic Nanomaterials: Synthesis, Characterization and Applications* (Cham: Springer International Publishing) 201–16

[34] Yao L and Xu S 2014 Detection of magnetic nanomaterials in molecular imaging and diagnosis applications *Nanotechnol. Rev.* **3** 247–68

[35] Issa B and Obaidat I 2019 Magnetic nanoparticles as MRI contrast agents *Magnetic Resonance Imaging* (London: IntechOpen) 4

[36] Mao X, Xu J and Cui H 2016 Functional nanoparticles for magnetic resonance imaging *Wiley Interdiscip. Rev.: Nanomed. Nanobiotechnol.* **8** 814–41

[37] Caspani S, Magalhães R, Araújo J and Sousa C 2020 Magnetic nanomaterials as contrast agents for MRI *Materials* **13** 2586

[38] Lapusan R, Borlan R and Focsan M 2024 Advancing MRI with magnetic nanoparticles: a comprehensive review of translational research and clinical trials *Nanoscale Adv.* **6** 2234–59

[39] Fatima H and Kim K 2018 Iron-based magnetic nanoparticles for magnetic resonance imaging *Adv. Powder Technol.* **29** 2678–85

[40] Mukhatov A, Le T, Pham T and Do T 2023 A comprehensive review on magnetic imaging techniques for biomedical applications *Nano Select* **4** 213–30

[41] Bushra R, Ahmad M, Alam K, Seidi F, Shakeel S, Song J, Jin Y and Xiao H 2024 Recent advances in magnetic nanoparticles: key applications, environmental insights, and future strategies *Sustain. Mater. Technol.* **40** e00985

[42] Oberdick S, Jordanova K, Lundstrom J, Parigi G, Poorman M, Zabow G and Keenan K 2023 Iron oxide nanoparticles as positive T1 contrast agents for low-field magnetic resonance imaging at 64 mT *Sci. Rep.* **13** 11520

[43] Zeinali Sehrig F, Majidi S, Asvadi S, Hsanzadeh A, Rasta S, Emamverdy M, Akbarzadeh J, Jahangiri S, Farahkhiz S and Akbarzadeh A 2016 An update on clinical applications of magnetic nanoparticles for increasing the resolution of magnetic resonance imaging *Artif. Cells, Nanomed. Biotechnol.* **44** 1583–8

[44] Aigbe U, Onyancha R U K, Okundaye B, Aigbe E, Enaroseha O, Obodo K, Osibote O, El Nemr A, Noto L and Atagana H 2023 Utility of magnetic nanomaterials for theranostic nanomedicine *Magnetic Nanomaterials: Synthesis, Characterization and Applications* (Cham: Springer International Publishing) 47–86

[45] Wadajkar A, Menon J, Kadapure T, Tran R, Yang J and Nguyen K T 2013 Design and application of magnetic-based theranostic nanoparticle systems *Recent Patents Biomed. Eng.* **6** 47–57

[46] Yadollahpour A, Hosseini S, Rashidi S and Farhadi F 2016 Applications of magnetic nanoparticles as contrast agents in MRI: recent advances and clinical challenges *Int. J. Pharm. Res. Allied Sci.* **5** 251–7 https://ijpras.com/article/applications-of-magnetic-nano-particles-as-contrast-agents-in-mri-recent-advances-and-clinical-challenges

[47] Yadollahpour A, Asl H and Rashidi S 2017 Applications of nanoparticles in magnetic resonance imaging: a comprehensive review *Asian J. Pharm.* **11** S7

[48] Yallapu M, Foy S, Jain T and Labhasetwar V 2010 PEG-functionalized magnetic nanoparticles for drug delivery and magnetic resonance imaging applications *Pharm. Res.* **27** 2283–95

[49] Neto D *et al* 2021 A novel amino phosphonate-coated magnetic nanoparticle as MRI contrast agent *Appl. Surf. Sci.* **543** 148824

[50] Hachani R *et al* 2017 Assessing cell-nanoparticle interactions by high content imaging of biocompatible iron oxide nanoparticles as potential contrast agents for magnetic resonance imaging *Sci. Rep.* **7** 7850

[51] Heydari Sheikh Hossein H, Jabbari I, Zarepour A, Zarrabi A, Ashrafizadeh M, Taherian A and Makvandi P 2020 Functionalization of magnetic nanoparticles by folate as potential MRI contrast agent for breast cancer diagnostics *Molecules* **25** 4053

[52] Scialla S, Genicio N, Brito B, Florek-Wojciechowska M, Stasiuk G, Kruk D, Bañobre-López M and Gallo J 2022 Insights into the effect of magnetic confinement on the performance of magnetic nanocomposites in magnetic hyperthermia and magnetic resonance imaging *ACS Appl. Nano Mater.* **5** 16462–74

[53] Ma K *et al* 2022 Using gradient magnetic fields to control the size and uniformity of iron oxide nanoparticles for magnetic resonance imaging *ACS Appl. Nano Mater.* **5** 7410–7

[54] Wabler M, Zhu W, Hedayati M, Attaluri A, Zhou H, Mihalic J, Geyh A, DeWeese T, Ivkov R and Artemov D 2014 Magnetic resonance imaging contrast of iron oxide nanoparticles developed for hyperthermia is dominated by iron content *Int. J. Hyperth.* **30** 192–200

IOP Publishing

Environmental Applications of Magnetic Nanomaterials
Under the Influence of Magnetic Fields

Uyiosa Osagie Aigbe and Kingsley Eghonghon Ukhurebor

Chapter 4

Applications of magnetic fields for environmental sustainability using magnetic nanomaterials

Since magnetic nanomaterials (MNMs) or magnetic nanoparticles (MNPs) have such exceptional nanoeffects and magnetic properties, they are frequently employed in environmental pollution detection. In view of these inherent qualities, MNMs or MNPs are being used for a wider range of environmental contamination detection applications. Consequently, this chapter will attempt to highlight the applications of magnetic fields (MFs) for environmental sustainability using MNMs or MNPs. In summary, current research on the use of MNMs or MNPs to identify organic and inorganic pollutants in soil and water, along with potential future applications, is discussed in this chapter.

4.1 Introduction

With regard to their special qualities, nanomaterials (NMs) or nanoparticles (NPs) have been attracting a lot of interest since their discovery. MNNs or MNPs are defined as magnetic particles with sizes of 0.1–100 nm, high surface chemical activity, and unique NM characteristics [1–5]. MNMs exhibit many characteristics, such as excellent dispersion, large specific surface area, and small particle size [6–8]. Furthermore, surface modification and copolymerization are two additional ways to alter MNMs [9]. The modification or functionalization of NMs for the purpose of detecting contaminants in the environment has garnered significant attention in recent years [10]. In the field of adsorbent NMs for wastewater pollution detection, MNMs are the most frequently employed type of NM. They can be utilized to absorb heavy metals (HMs), inorganic salt pollutants, and trace organic pollutants [11, 12].

Due to these characteristics and qualities, MNMs are more effective at detecting pollutants. MNMs have a modest magnetic effect because of their small particle sizes. MNMs' activity is increased by their ability to absorb more of the chemical under test due to their large specific surface area. Effective separation of MNMs in the presence of an external magnetic field (MF) is made possible by their considerable magnetic responsiveness [13, 14]. At the moment, ferric oxide (Fe_2O_3) NMs [15, 16], magnetite (Fe_3O_4) [17, 18], and nano zero-valent iron (nZVI) are the most commonly employed MNMs [19, 20]. The basic idea behind MNM detection is that the material to be evaluated and measured can easily be removed from the sample by interacting and binding with MNMs under specific conditions; then, the material may be separated, identified, and detected using an external MF [21].

The application of MNMs in the identification of soil and water pollution has garnered significant interest [22–26]. A current area of study focuses on modifying MNMs to more effectively enrich particular contaminants or measuring the biological toxicity of samples directly utilizing microbial composition biosensors (BSs)/nanobiosensors (NBSs) and the properties of MNMs [27–29]. Many published studies have discussed the development, modification, and other uses of MNMs. Nevertheless, comprehensive and extensive publications on the use of MNMs in environmental detection, especially on the applications of MFs and MNMs for environmental sustainability, are still insufficient. The many different methods for synthesizing MNMs with various characteristics are included in chapter two of this book. In the meantime, instances of MNM application in environmental detection in soil and water are discussed, since the concepts and techniques of their functional applications were already described in chapter two of this book. Hence, this chapter discusses the advancements and benefits of MNMs or MNPs in environmental detection vis-à-vis the applications of MFs for environmental sustainability using MNMs or MNPs, along with potential future applications.

4.2 Applications of MNMs for environmental detection

Toxins such as HM ions, pesticides, dyes, and other chemical toxins, especially those from industrial activities, pose a serious threat to the ecosystem since they are found in water, land, food, and the atmosphere [30–33]. These environmental issues can be addressed through the use of MNMs [34–39], due to their properties, including facile surface encapsulation, high stability, and separability [40–42]. MNMs have been employed extensively. Enriching compounds that pollute the environment and merging with bacteria to create BSs/NBSs for detecting environmental standards are two of the primary applications for MNMs in environmental detection. MNMs are currently widely employed in soil and water detection [43–45]. Accordingly, some of the environmental samples that have been detected utilizing MNMs are contained in table 4.1, as reported by Zhang *et al* [46].

4.2.1 Applications of MNMs for detection in aquatic environments

MNMs are frequently employed to detect different types of contaminants within water (aquatic) environments, usually in conjunction with additives [2–5, 50–52].

Table 4.1. Analytical techniques or methods comparison with various environmental sample categories.

Categories of MNMs or MNPs	Categories of analyte/pollutant	Sample type	Analytical methods	Recovery (%)	LOD	Authors
Gn-MNMs	SR dye	Water	Adsorption/desorption	98.12–103.52	1.80–5.50 ng l^{-1}	Wu et al [47]
Fe$_3$O$_4$ or Al$_2$O$_3$ NMs	Sulfonamides	Soil	Extraction, concentration, separation	71.00–93.00	0.37–6.74 ng g^{-1}	Sun et al [48]
Fe$_3$O$_4$ NMs	N-(phenylmethyl)-9H-purin-6-amine	Food	Static adsorption	82.63–106.27	150.00 ng ml^{-1}	Cao et al [49]

Notes: LOD = limit of detection, SR = Sudan red.

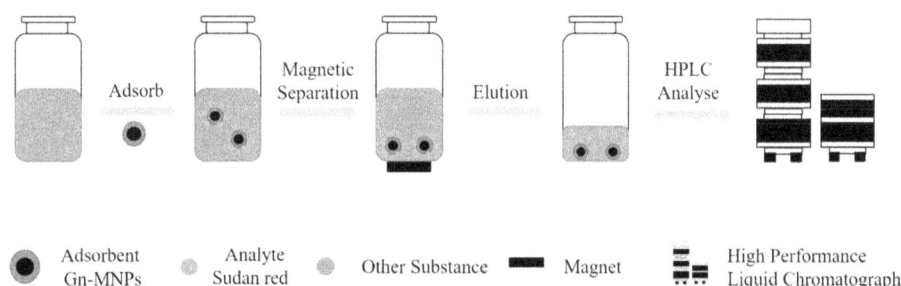

Figure 4.1. Representative illustration of the Gn-MNM or Gn-MNP procedure. Reproduced from [47]. CC BY 4.0.

The two primary categories of contaminants found in water are organic and inorganic. Organic materials encompass a variety of substances, such as chemicals [53], medicines and pharmacological products [54–56], dyes [47, 57, 58], and insecticides [59]. Acids and HM ions, among others, make up the majority of inorganic compounds [60–63].

Wu et al [47] synthesized, developed, and studied a novel category of MNMs known as polyamide amine dendrimer-modified MNMs or MNPs (Gn-MNMs or Gn-MNPs), as shown in figure 4.1 (adapted and reproduced from Wu et al [47]). The SR dye in the solution was absorbed by stirring for sixty minutes while the adsorbent Gn-MNM or Gn-MNP was at work. Through the use of magnetism, the adsorbent and original solution were separated. Acetone was then used to remove the SR dye from the adsorbent. The residue was eluted with methanol (CH$_3$OH). After the resulting solution was dried with nitrogen, the residue was disseminated. Random samples were collected and sent for analysis. In natural water, the NM exhibited good adsorption performance for dyes such as Congo red, SR, and methyl green. The limit of detection (LOD) for Congo red was between 1.80 and 5.50 ng l^{-1}. With a precision of below 3.00%, the material offers a wide range of applications in the concentration of trace contaminants from environmental water samples. This NM has great potential for use in concentrating trace contaminants in environmental water samples.

MNMs have also been employed in environmental water sample enrichment and pesticide detection. When added to samples, the recovery rate can reach 90% in real-world situations. A cupric organic skeleton/ferric oxide (MOF–199/Fe_3O_4) complex was produced *in situ* at room temperature by Cao *et al* [49] and subsequently employed for neonicotinoid insecticide identification and extraction in ambient water samples. Using molecular imprinting technology, Turiel *et al* [64] altered the surface of MNMs and investigated their efficacy as a selective adsorbent for removing triazine chemicals from ambient water. Furthermore, Kouhestani and Ebrahimi [59] synthesized MNMs coated with chitosan functionalized with cysteine, and they employed this material as an efficient adsorbent to recover carboxin and bensulfuron sulfomethyl. Cui *et al* [65] synthesized magnetic molecularly imprinted particles (MMIPs) by nucleating MNPs of Fe_3O_4 and polymerizing molecularly imprinted polymers using imidacloprid as the starting material and dopamine as a functional monomer. Imidacloprid was obtained from water samples using MMIPs and high-performance liquid chromatography (HPLC) under optimal conditions.

Drugs present in water may additionally be concentrated by MNMs. Ragab and Bahgat [66] produced MNMs enhanced with cetyltrimethylammonium bromide (CTAB) and subsequently performed preliminary enrichment, detection, and solid-phase extraction of valsartan in water samples. This technique's basic idea is to adsorb the chemical under test using MNMs and then utilize CH_3OH for desorption. For the purpose of separating and desorbing the material from the solution, a mobile phase made of phosphate buffer (0.03 M), acetonitrile, and CH_3OH (40:40:20%) was utilized. Furthermore, the separated solution was subjected to chromatographic analysis to determine the substance's concentration for testing. Throughout a concentration range of 10.00–150.00 ng ml^{-1}, these MNMs exhibited an excellent linear response, and the LOD was as low as 2.02 ng ml^{-1}.

MNMs are frequently utilized in the identification of organic reagents because they are fast at extracting the material to be tested from the extraction solution. Wu *et al* [67] produced Fe_3O_4 MNMs that were dendrimer-modified by polyamides. The modified MNMs were employed as efficient adsorbents in the magnetic solid-phase extraction (MSPE) of trace tetrabromobisphenol A and 4-nonylphenol from ambient water samples. Following this procedure, bisphenol A-containing extraction solvent was extracted more easily with the help of MNMs. Simultaneously, Tian *et al* [53] synthesized $Fe_3O_4@SiO_2@CTS$, a modified natural polysaccharide chitosan (CTS), on the surface of magnetic Fe_3O_4 MNs for the purpose of detecting polychlorinated biphenyls (PCBS). These findings suggest that MNMs are reusable and recyclable, exhibit facile magnetic separation, and have strong selectivity.

Additionally, there are numerous novel applications of nanotechnology for the identification and mitigation of water pollution. A new sulphur (S)-doped carbon nitride (CN)/reduced graphene oxide (rGO) porous nanosheet (S-CN/rGO PNs) was synthesized by Zheng *et al* [68] via supramolecular self-assembly and solvothermal treatment. Compared to regular materials, the photocatalytic action of Cr(VI) and Rhodamine B increased seventeen times under the influence of this substance.

Chitosan–PANI–Fe_2O_3 was developed by Singh et al [69] using a batch adsorption method. With a removal efficacy of 91.50%, this composite adsorbed and removed colors/dyes such as methyl orange from water. Furthermore, Khan et al [70] provided a new method for water treatment by synthesizing research into 3D-printed nanosheet membranes for water management and purification.

MSPE uses magnetic or magnetized materials as adsorbent substrates and is based on dispersible solid-phase extraction (SPE) technology [71–73]. NMs can achieve smaller microextraction concentrations, outstanding extraction capability, and more extraction efficiency compared to conventional SPE fillers. NMs require less adsorbent and shorter equilibrium times because these particles possess a greater specific surface area as well as a smaller diffusion distance. According to Zhang et al [6], the following are the benefits of this technique: a good preliminary enrichment factor, quick extraction speed, accurate detection, and low relative variation in standard deviation (SD).

To capture and detect HMs such as Hg^{2+}, Kong et al [74] synthesized manganese-based MNMs using the microemulsion technique, covered the surface with gold (Au), and bound cysteine through the Au–S covalent connection. As a result of their exceptional efficacy as HM nanoadsorbents, these MNMs helped determine the concentration of Hg^{2+} ions in aqueous solutions used in environmental management and evaluations. The magnetic adsorbent exhibited rapid adsorption kinetics, substantial adsorption capacity, and strong Pb^{2+} selectivity. Zhao et al [73] developed double-imprinted polymer-coated MNMs with template Pb^{2+} ion co-imprinting using 4.00 nm ZnO NMs as a sacrificial template. Graphite furnace atomic absorption spectrometry was used to detect trace Pb^{2+}. Jagirani et al [51] used magnetic cellulose NMs (Cell-MNMs) as an adsorbent to recover trace lead from environmental materials using solid-phase microextraction (SPME). Cell-MNMs were effectively employed for the SPE of Pb^{2+} in water samples by adjusting the analytical settings. To detect Cr^{3+}, Ni^{2+}, Co^{2+}, and Hg^{2+} in water from environmental samples, Wu et al [52] employed a magnetic solid Schiff base as a multifaceted agent, acetonitrile/water (60:40, v/v) as a dispersive solvent, and 1-octanol as an extractant for dispersive liquid–liquid microextraction using hydrophilic MNMs for liquid-phase extraction of 1-octanol.

Second-generation amino-dendrimer functionalized MNMs ($Fe_3O_4@G2$-PAD) were produced by Maleki et al [75] and used to monitor Pb^{2+} and Cd^{2+} ions in aquatic environments (such as lakes, rivers, and wastewater). Wu et al [76] employed Fe_3O_4 MNMs and glutathione-modified silver (Ag) NMs (AgNMs) for the selective trace detection of Pb^{2+} and Cd^{2+}. AgNM aggregation was induced by Pb^{2+} and Cd^{2+} through co-metal ligand interactions, and the resulting aggregate was loaded onto magnetic Fe_3O_4 NMs as an adsorbent. At a concentration of $10.00 \, g \, l^{-1}$, the average SDs for Cd and Pb were 1.5% and 1.8%, respectively, and their corresponding LODs were 0.13 and $1.25 \, g \, l^{-1}$. In order to investigate the sensitivity of a surface-enhanced Raman scattering (SERS) substrate for the quick monitoring and detection of Cr^{6+}, Wang et al [77] utilized magnetic $Fe_3O_4/ZrO_2/Ag$ composite microspheres. They then adopted the optimized $Fe_3O_4/ZrO_2/Ag$ system for the quantitative monitoring and detection of Cr^{6+} in an aqueous solution. Their findings demonstrated a strong linear correlation ($R^2 = 0.98$) between the logarithmic concentration of Cr^{6+} and the SERS intensity, with a potential LOD of as low as 10^{-7} M. The inductively coupled plasma mass spectrometry (ICP-MS) technique was employed to identify the sulfhydryl-amino functionalized NMMs $Fe_3O_4@SiO_2@MPTMS$ and $Fe_3O_4@SiO_2@APTES$, which were utilized as adsorbents in MSPE for the direct extraction of As(III) and As(V), respectively [78]. Coordination and electrostatic forces were used to recover the amino extract As(V) and sulfhydryl extract As(III), respectively. The recovery rates for As(III) and As(V) were 89.00%–96.00% and 90.00%–102.00%, respectively. These techniques produced good results when used to assess the amounts of HMs in water samples collected from the environment [6].

HMs in water can be detected both qualitatively and quantitatively using colorimetric or optical techniques combined with MNMs. These techniques have the benefits of good selectivity, outstanding stability, and ease of use [79, 80]. Oguz *et al* [81] synthesized Fe_3O_4 NMs from fluorescent chemicals and used the luminescent intensity of the MNMs in water to identify Hg^{2+} in a water-based environment. AuNMs were synthesized by Liu *et al* [82] and applied to an Fe_3O_4 NM surface to enable the optical identification of $Au@Fe_3O_4$ NMs with elevated Hg^{2+} concentrations in pollutants from industrial processes. This technique can be qualitatively utilized for the analysis and monitoring of Hg^{2+} beyond 5.00 μM by the unaided eye and has exceptionally high sensitivity and selectivity. The fluorescent approach is a useful tool for pollution detection. Magnetic $Fe_3O_4@SiO_2$-TbDPA nanoprobes were synthesized by Li *et al* [83]. A water-based solution of synthetic $Fe_3O_4@SiO_2$-TbDPA showed significant green fluorescence, the magnitude of which decreased when NO^{2-} (0–100 μM) was introduced at various concentrations. It was possible to determine the corresponding NO^{2-} ion quantity in the solution under analysis by referring to a premeasured inhibitory luminescence curve. The broad applicability of this method in both environmental management and biomedical industries is made possible by its good linearity in the concentration range of 5–80 μM and its LOD of 1.03 μM.

Extensive studies have been conducted on MNMs as a means of detecting various types of harmful substances in water [6]. The results of these studies have been promising. Nevertheless, most modified MNMs are difficult to prepare and may not be reused [6]. MNMs that are readily produced and reusable could potentially be developed in the future.

4.2.2 Applications of MNMs for detection in soils

HMs and organic materials in contaminated soil have broad biological toxicity that can endanger the health of humans, poison people, and even have an adverse effect on the natural environment [84, 85]. Since soil biotoxicity has received so much consideration, research has focused on accurately detecting soil biotoxicity when different kinds and forms of contaminants are present [86]. MNMs have found extensive applications in soil environmental monitoring in the last few years, owing to their rapid adsorption rate and facile magnetic separation [6].

MNMs have the primary benefit of being readily and effectively extracted from the processed solution with a standard magnet. Using chemical coprecipitation,

Singh *et al* [24] produced Fe_3O_4 NMs, which were then utilized to extract HMs such as Cd, Cr, Cu, Fe, Ni, Pb, and Zn from soil samples. Between 69.60% and 99.60% of HMs were adsorbed when the pH of the soil leaching solution was 0.7. Atomic absorption spectrometry was used to measure the adsorption efficiency of MNMs for different HMs. The strongest adsorption and greatest detection effectiveness of the MNMs were observed for Pb in soil, based on the data.

In order to identify the presence of Cs in soil and selectively separate Cs-polluted clay composites from soil, Kim *et al* [87] employed bare Fe_3O_4 NMs. The Cs recovery rate exceeded 90.00% at a low pH and a mass ratio of MNMs to clay of about 0.10. The explanation is that at low pH values, the magnetic separation of clay materials is facilitated by an increase in their electrostatic force and dispersion. Using MNMs, Kasa *et al* [88] developed a quick and efficient procedure for dispersive SPME of Pd in soil. Pd was extracted directly from the soil sample solution without requiring complexation. The method utilized was slotted quartz tube flame atomic absorption spectrometry (SQT-FAAS). Using this method, Pd was quantified to $21.40 \, ng \, ml^{-1}$ and detected at $6.40 \, ng \, ml^{-1}$, with a relative standard deviation of 6.60%. This technique's recoveries, which range from 90% to 101%, demonstrate its precision and practicality.

Modified MNMs are being used by some researchers to identify and remove pesticides from soil [6]. To perform SPE of triazine in soil samples, Patiño-Ropero *et al* [89] produced a promazine-imprinted polymer on the surface of modified MNMs. Without the need for extra centrifugation or filtration processes, the produced MNMs had high selectivity for triazine composites and were easy to collect and separate using an external MF. Depending on the triazines and the kind of soil employed, the recovery rate varied from 5.40% to 40.60% with a relative SD of less than 7.00% ($n = 3$) and an LOD of $0.10–3.00 \, ng \cdot g^{-1}$.

In order to analyze sulfonamides using magnetic SPE of sulfadimethoxine and sulfaquinoxaline in various soil samples, Sun *et al* [48] produced alumina-coated magnetite NMs (Fe_3O_4/Al_2O_3 NMs). By combining an extraction solvent and a magnetic adsorbent with a soil sample while an ultrasonic wave is operating, the extraction and concentration process can be completed in a single step. The sulfonamides desorbed from the adsorbent are subsequently identified by liquid chromatography–tandem mass spectrometry. The adsorbent is then readily extracted from the multifaceted substrate by an applied MF. In order to create a hybrid nanomagnetic 1-Fe_3O_4, Kim *et al* [90] attached naphthalimide DPA(2) to the surface of Fe_3O_4 NMs. At 527 nm, there was a noticeable rise in fluorescent intensity due to the binding of 1-Fe_3O_4 and naphthalimide DPA(2) to Zn^{2+}. Zn^{2+} was selectively detected and removed from soil samples using the nanomagnetite 1-Fe_3O_4. Before performing an HPLC analysis, Yang *et al* [91] produced a polyionic liquid (PIL), fixed it on SiO_2-coated MNMs that had been manufactured, and then utilized the PIL as an adsorbent in MSPE to extract sulfonylurea herbicides (SUHs) from soil samples. Analytes used in repeated assays had relative SDs between 3.20% and 4.50%. The sensitivity and LOD ranges were $5.4–9.8 \, ng \cdot ml^{-1}$ and $1.62–2.94$ $ng \cdot ml^{-1}$, respectively. These findings suggest that sulfonylurea herbicides can be efficiently extracted from soil samples using $Fe_3O_4@SiO_2@PIL$.

Pang *et al* [92, 93] employed MNMs in contaminated soil to enrich and identify HM ions and sulfonylurea herbicides. They introduced doped Fe_3O_4 MNMs into a polyethylene imidazole dimethyl acrylate polymer monolithic capillary microextraction column (MCMC). SUHs can be concentrated with an MF to achieve extraction rates ranging from 82.60% to 94.50%. This method identified the presence of trace SUHs with a LOD range (S/N = 3.00) of 0.30–1.50 $\mu g\,kg^{-1}$ in soil samples. In addition, Cr^{3+}/APD and Cr^{6+}/APD complexes were formed by coordinating Cr^{3+} and Cr^{6+} with pyrrolidine ammonium dithiocarbamate (APD). Subsequently, in a capillary tube, a porous monomer microextraction column doped with MNMs was produced in real time. The findings demonstrated that the extraction rates of Cr^{3+}/APD (80.4%) and Cr^{6+}/APD (86.20%) complexes were enhanced by the application of an MF. In soil samples, the LODs for Cr^{3+} and Cr^{6+} were 0.47 and 0.057 $\mu g\,kg^{-1}$, respectively. Cr^{3+} and Cr^{6+} with enrichment factors of 59 and 72, respectively. Li *et al* [94] used the solvothermal approach to produce MOF-1210 (Zr/Cu)-modified MNMs in a single step for the MSPE of benzophenone. The extraction and identification of benzophenone from soil samples were effectively accomplished using this approach; the relative standard deviation (RSD) was less than 11.12%, and recovery rates ranged from 87.60% to 113.80%. In order to measure the residues of pesticides in soil samples from agricultural locations, Hubeyska *et al* [95] developed the QuEChERS purification and extraction procedure, which is based on hydrophobic MNMs (C18/GCB/Fe_3O_4/Fe_3O_4@Triton). They subsequently combined this technique with mass spectrometry–gas chromatography for the simultaneous identification of 16 organochlorine pesticides. Kim *et al* [96] employed Fe_3O_4 NMs coated in polyethylenimide (PEI) to separate particles of clay from soil polluted with Cs in a selective manner. Through electrostatic attraction, the PEI coated on the surface of nano-Fe_3O_4 increased the interaction force between the magnetic nano-Fe_3O_4 and the clay minerals. At a low concentration of 0.04 NM/g clay, the Fe_3O_4-PEI NM magnetically extracted almost all of the particles of clay from the solution.

In comparison to water-based approaches, MNMs that are applied to adsorb contaminants directly in soil exhibit weaker adsorption as well as lower rates of recovery. The reason for this is that MNMs are challenging to extract efficiently from soil. As a result, researchers will need to explore MNMs with greater recovery efficiencies in soil, as well as novel applications that would be beneficial there [6].

Accurately detecting soil biotoxicity is the first step—and sometimes the last—in the assessment and restoration of contaminated soil in environmental assessments of soil [97]. The process of determining the level of contaminants and evaluating their biotoxicity via models by means of analytical chemical detection technologies has encountered issues in the last few years [6, 98]. Chemical agents and centrifugal shock, as well as other types of detection, monitoring, and treatment, can damage both the chemical and physical attributes of contaminants when employing chemical extraction and other force-based techniques to extract pollutants. This makes it impossible to accurately determine the biological toxicity of polluted soil. Specifically, the model is frequently unable to precisely and quantitatively characterize the overall biological toxicity when several contaminants exhibit susceptibility and synergistic impacts [6]. Nevertheless, rather than being an immediate indication of toxicity, the identification of biological toxicity by conventional analytical chemical procedures is simply an assessment of contaminant content. The microbial detection approach, which is seen as an extremely promising technique for the detection of biotoxicity in contaminated soil, is a straightforward identification of the biotoxicity of detrimental substances transferred to the liquid phase in soil contamination as opposed to the analytical chemical method. Microorganisms can be coupled with MNMs manufactured using biofriendly procedures, modified, and functionalized to form BSs. This can effectively be used to build a microbial monitoring and detection method for MNM-based BSs and NBSs. The complete biotoxicity of multiply contaminated soil may be precisely characterized using this very reproducible and sensitive approach [6].

A whole-cell bioreporter produced by Jia *et al* [99] integrated MNMs with genetically altered *Escherichia coli* to assess soil ecotoxicity based on luminescent intensity. In order to minimize the influence of particles of soil, permanent magnets may be used in conjunction with MNMs or MNPs to recover BSs or NBSs from soil samples. The sensor's operational mechanism is shown in figure 4.2, as adapted and reproduced from Jia *et al* [99]. The luminescent intensity can be measured and monitored using the MMS solution to assess soil ecotoxicity, and the BS or NBS is then removed from the solution using a magnetic probe once it has reacted in the soil diluent. When evaluating the toxicity and bioavailability of Cr contaminants in soil, a bioreporter functionalized with MNMs by a magnetic device (MFB) had greater sensitivity and improved reproducibility than the conventional treatment, which is based on directly adding a bioreporter to a mixture of soil and water (SW-M treatment) or a supernatant (SW-S treatment). Zhang *et al* [100] synthesized a magnetic nanobacterial sensor to detect soil biological toxicity by combining MNMs with a bright luminescent bacillus. This sensor's unique feature is that it can determine the biological toxicity of contamination in soil with greater accuracy, since the physical or chemical characteristics of the soil contaminants do not need to be changed [6, 101].

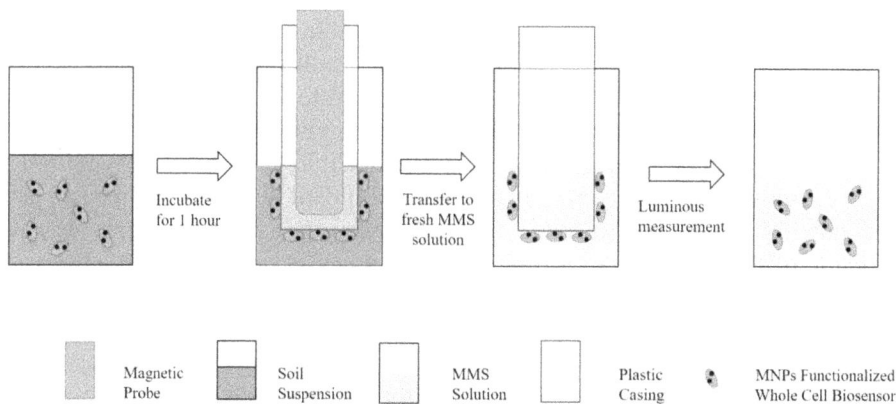

Figure 4.2. Representative illustration of the magnetic BS device produced by Jia *et al*. Reprinted from [99], Copyright (2016), with permission from Elsevier.

4.3 Conclusions and potential future applications of MNMs for environmental detection

This chapter has highlighted the research progress and advancements of MNMs in environmental detection and their applications for environmental sustainability in conjunction with MFs. MNMs and their possible applications in environmental remediation and sustainability are outlined in this chapter. MNMs are useful for the remediation of the environment due to their quick removal of biological contaminants and pollutants. Given that MNMs have unique optical, magnetic, thermal, and mechanical properties, they are being used to explore novel applications.

Due to MNMs' ability to convert magnetic energy into thermal energy, a plethora of analytical techniques is now possible. MNMs are very effective due to their reusability and simple separation by magnets. Additionally, their use makes it possible to overcome the challenges associated with using adsorbent materials as well as the time and sample volume requirements. The utilization of MNMs and their mode of action in diverse environmental restoration and sustainability applications have been described in this chapter. Since more industries are releasing toxic substances into the environment, contamination of the environment has increased in recent years. To safely eliminate both organic and inorganic contaminants from water, innovative water treatment techniques based on the utilization of adsorptive materials like carbon and adsorbents are desperately needed. MNMs have the ability to extract organic and inorganic pollutants from water. MNMs are more precise and effective in concentrating pollutants than conventional detection techniques. Furthermore, it has been shown that the biological production of MNMs promotes sustainable development and is not harmful to the environment. According to some studies, MNMs have strong application prospects in the detection of soil contamination for environmental sustainability; nevertheless, further research is necessary to fully realize their popularization and utilization.

Although MNMs have performed well in environmental testing, there are still some challenges requiring attention, such as the challenging MNM preparation processes and the low rates of MNM recovery in soil. The biological toxicity of environmental samples may not be directly detectable by MNMs, despite the fact that they can directly remove contaminants for quantitative measurement from soil and water. This chapter concludes that the biological toxicity of materials related to the environment can be accurately measured by mixing MNMs with microbes to create an MNM–microbe sensor. Because this technique uses an external MF, it isolates the sensor from the contaminated sample and allows for direct measurement of the material's biological toxicity without the need for additional steps. However, the efficacy of synthetic sensors is unreliable because conventional MNMs have poor biological functions and easily aggregate. Today, there is strong research interest in the preparation and development of MNMs with stable morphologies, exceptional dispersion performance, excellent biological functioning and biocompatibility, excellent stability under operational conditions, as well as eco-friendliness and environmental sustainability. The most effective substitute technique for creating MNs in a way that is favorable to biology is biomimetic mineralization; however, this approach currently faces issues including high preparation costs and limited yield. To enable the broad use of biofriendly MNMs in sectors such as biomedicine, agriculture, and the environment, more research into biomimetic mineralization techniques needs to be conducted in the future.

References

[1] Gupta A and Gupta M 2025 Synthesis and surface engineering of iron oxide nanoparticles for biomedical applications *Biomaterials* **26** 3995–4021

[2] Singh R, Ukhurebor K, Singh J, Adetunji C and Singh K 2022 *Nanobiosensors for Environmental Monitoring - Fundamentals and Application* (Cham: Springer Nature)

[3] Aigbe U, Ukhurebor K and Onyancha R 2023 *Synthesis, Characterization and Applications of Magnetic Nanomaterials: Recent Advances* (Cham: Springer Nature)

[4] Ukhurebor K, Aigbe U and Onyancha R 2023 *Adsorption Applications for Environmental Sustainability* (Bristol: Institute of Physics (IOP) Publishing)

[5] Ukhurebor K and Aigbe U 2024 *Environmental Applications of Magnetic Sorbents* (Bristol: Institute of Physics (IOP) Publishing)

[6] Zhang K, Song X, Liu M, Chen M, Li J and Han J 2023 Review on the Use of Magnetic Nanoparticles in the Detection of Environmental Pollutants *Water* **15** 3077

[7] Bushra R, Ahmad M, Alam K, Farzad Seidi F, Qurtulen, Shakeel S, Song J, Jin Y and Xiao H 2024 Recent advances in magnetic nanoparticles: key applications, environmental insights, and future strategies *Sustain. Mater. Technol.* **40** e00985

[8] Zhang Y, Cheng J and Liu W 2019 Characterization and relaxation properties of a series of Monodispersed magnetic nanoparticles *Sensors* **19** 3396

[9] Zhu N, Ji H, Yu P, Niu J, Faroop M, Akram M, Udego I, Li H and Niu X 2018 Surface modification of magnetic iron oxide nanoparticles *Nanomaterials* **8** 810

[10] Lu R, Wang C, Chen Y, Tan L, Wang P and Feng S 2022 IL-functionalized Mn(II)-doped core–shell Fe3O4@Zr-MOF nanomaterials for the removal of MB from wastewater based on dual adsorption/Fenton catalysis *New J. Chem.* **46** 8534–44

[11] Chai Z, Yuan C, Qian Y, Zhang Y and Wang J 2019 Developed on the Adsorption Properties of 9 Organophosphorus Pesticides in Spinach by Magnetic Nanoparticles Modified Carbon Nanotube Composites *Food Ferment. Sci. Technol.* **54** 70–5

[12] Lou X, Boada R, Verdugo V, Simonelli L, Pérez G and Valiente M 2022 Decoupling the adsorption mechanisms of arsenate at molecular level on modified cube-shaped sponge loaded superparamagnetic iron oxide nanoparticles *J. Environ. Sci.* **121** 1–12

[13] Yang J, Zhou L, Ma F, Zhao H and Deng F 2020 S. Pi and A. Tang, 'Magnetic nanocomposite microbial extracellular polymeric substances@Fe3O4 supported nZVI for Sb (V) reduction and adsorption under aerobic and anaerobic conditions *Environ. Res.* **189** 109950

[14] Plastiras O-E, Deliyanni E and Samanidou V 2021 Synthesis and Application of the Magnetic Nanocomposite GO-Chm for the Extraction of Benzodiazepines from Surface Water Samples Prior to HPLC-PDA Analysis *Appl. Sci.* **11** 7828

[15] Yu L, Peng X, Ni F, Li J, Wang D and Luan Z 2013 Arsenite removal from aqueous solutions by γ-Fe2O3-TiO2 magnetic nanoparticles through simultaneous photocatalytic oxidation and adsorption *J. Hazard. Mater.* **2013** 246–7

[16] Akhbarizadeh R, Shayestefar M and Darezereshki E 2014 Competitive Removal of Metals from Wastewater by Maghemite Nanoparticles: A Comparison Between Simulated Wastewater and AMD *Mine Water Environ.* **33** 89–96

[17] Al-Qasmi N, Almughem F, Jarallah S and Almaabadi A 2022 Efficient Green Synthesis of (Fe3O4) and (NiFe2O4) Nanoparticles Using Star Anise (Illicium verum) Extract and Their Biomedical Activity against Some Cancer Cells *Materials* **15** 4832

[18] Nguyen M, Tran H-V, Xu S and Lee T 2021 Fe3O4 Nanoparticles: Structures Synthesis, Magnetic Properties, Surface Functionalization, and Emerging Applications *Appl. Sci.* **11** 11301

[19] Zhou Z, Huang J, Xu Z, Ali M, Shan A, Fu R and Lyu S 2021 Mechanism of contaminants degradation in aqueous solution by persulfate in different Fe(II)-based synergistic activation environments: Taking chlorinated organic compounds and benzene series as the targets *Sep. Purif. Technol.* **273** 118900–98

[20] Vilardi G, Rodríguez-Rodríguez J, Ochando-Pulido J, Verdone N, Martinez-Ferez A and Palma L 2018 Large Laboratory-Plant application for the treatment of a Tannery waste-water by Fenton oxidation: Fe(II) and nZVI catalysts comparison and kinetic modelling *Process Saf. Environ. Prot.* **117** 629–38

[21] Wierucka M and Biziuk M 2014 Application of magnetic nanoparticles for magnetic solid-phase extraction in preparing biological, environmental and food samples *Trends Anal. Chem.* **59** 50–8

[22] Kim J-H, Kim S-M, Yoon I-H, Choi S-J and Kim I 2020 Selective separation of Cs-contaminated clay from soil using polyethylenimine-coated magnetic nanoparticles *Sci. Total Environ.* **136020** 136020

[23] Li T, Yu Z, Yang T, Xu G, Guan Y and Guo C 2021 Modified Fe_3O_4 magnetic nanoparticles for COD removal in oil field produced water and regeneration *Environ. Technol. Innov* **23** 101630

[24] Singh A, Chaudhary S and Dehiya B 2021 Fast removal of heavy metals from water and soil samples using magnetic Fe_3O_4 nanoparticles *Environ. Sci. Pollut. Res.* **28** 3942–52

[25] Ukhurebor K, Hossain I, Pal K, Jokthan G, Osang F, Ebrima F and Katal D 2023 Applications and contemporary issues with adsorption for water monitoring and remediation: a facile review *Top. Catal.* **67** 140–55

[26] Anani A, Adama K, Ukhurebor K, Habib A, Abanihi V and Pal K 2023 Application of nanofibrous protein for the purification of contaminated water as a next generational sorption technology: a review *Nanotechnology* **34** 232004

[27] Heo Y, Lee E and Lee S-W 2022 Adsorptive removal of micron-sized polystyrene particles using magnetic iron oxide nanoparticles *Chemosphere* **307** 135672

[28] Guo M, Huang K and Xu W 2021 Third generation whole-cell sensing systems: Synthetic biology inside, nanomaterial outside *Trends Biotechnol.* **19** 550–9

[29] Bushra R, Ahmad M, Alam K, Farzad Seidi F, Shakeel Qurtulen S, Song J, Jin Y and Huining Xiao H 2024 Recent advances in magnetic nanoparticles: key applications, environmental insights, and future strategies *Sustain. Mater. Technol.* **40** e00985

[30] Ukhurebor K, Aigbe U, Onyancha R, Nwankwo W, Osibote O, Paumo H, Ama O, Adetunji C and Siloko I 2021 Effect of hexavalent chromium on the environment and removal techniques: a review *J. Environ. Manage.* **280** 111809

[31] Kerry R *et al* 2021 A comprehensive review on the applications of nano-biosensor based approaches for non-communicable and communicable disease detection *Biomaterials* **9** 3576–602

[32] Paladhi A, Manohar M, Pal K, Vallinayagam S, Packirisamy A, Bashreer V, S.R N and Ukhurebor K 2022 Novel Electrochemical biosensor key significance of smart intelligence (IoMT & IoHT) of COVID-19 virus control management *Process Biochem.* **122** 105–9

[33] Onyancha R, Ukhurebor K, Aigbe U, Osibote O, Kusuma H, Darmokoesoemo H and Balogun V 2021 A systematic review on the detection and monitoring of toxic gases using carbon nanotube-based biosensors *Sens. Bio-Sens. Res.* **34** 100463

[34] Aidonojie P, Ukhurebor K, Oaihimire I, Ngonso B, Egielewa P, Akinsehinde B, Heri S and Darmokoesoemo H 2023 Bioenergy revamping and complimenting the global environmental legal framework on the reduction of waste materials: a facile *Heliyon* **9** e12860

[35] Aigbe U, Ukhurebor K, Onyancha R, Osibote O, Kusuma H and Darmokoeso H 2022 Measuring the velocity profile of spinning particles and its impact on Cr(VI) sequestration *Chem. Eng. Process.* **178** 1–15

[36] Aigbe U, Ukhurebor K, Onyancha R, Okundaye B, Pal K, Osibote O, Esiekpe E, Kusuma H and Darmokoesoemo H 2022 A facile review on the sorption of heavy metals and dyes using bionanocomposites *Adsorpt. Sci. Technol.* **2022** 8030175

[37] Ukhurebor K, Onyancha R, Aigbe U, UK-Eghonghon G, Kerry R, Kusuma H, Darmokoesoemo H, Osibote O and Balogun V 2022 A methodical review on the applications and potentialities of using nanobiosensors for diseases diagnosis *BioMed Res. Int.* **2022** 1682502

[38] Aigbe U, Ukhurebor K, Onyancha R, Osibote O, Darmokoesoemo H and Kusuma H 2021 Fly ash-based adsorbent for adsorption of heavy metals and dyes from aqueous solution: a review *J. Mater. Res. Technol.* **14** 2751–74

[39] Onyancha R, Aigbe U, Ukhurebor K and Muchiri P 2021 Facile synthesis and applications of carbon nanotubes in heavy-metal remediation and biomedical fields: a comprehensive review *J. Mol. Struct.* **1238** 130462

[40] Lian L, Zhang X, Hao J, Lv J, Wang X, Zhu B and Lou D 2018 Magnetic solid-phase extraction of fluoroquinolones from water samples using titanium-based metal-organic framework functionalized magnetic microspheres *J. Chromatogr. A* **1579** 1–8

[41] Lu D, Qin M, Liu C, Deng J, Shi G and Zhou T 2021 Ionic liquid-functionalized magnetic metal-organic framework nanocomposites for efficient extraction and sensitive detection of fluoroquinolone antibiotics in environmental water *ACS Appl. Mater. Interface* **13** 5357–67

[42] Serantes D, Chantrell R, Gavilán H, Morales M, Chubykalo-Fesenko O, Baldomir D and Satoh A 2018 Anisotropic magnetic nanoparticles for biomedicine: bridging frequency separated AC-field controlled domains of actuation *Phys. Chem. Chem. Phys.* **20** 30445–54

[43] Abu-Dief A, Abdelbaky M, Martínez-Blanco D, Amghouz Z and García-Granda S 2016 Effect of chromium substitution on the structural and magnetic properties of nanocrystalline zinc ferrite *Mater. Chem. Phys.* **174** 164–71

[44] He Y, Wang Y, Zhao W, Gao R and Lv D 2021 Preparation of magnetic nano-composite adsorbents and their application in water treatment *Ind. Water Treat* **41** 1–6

[45] Li M, Hu L, Niu C, Huang D and Zeng G 2018 A magnetic separation fluorescent aptasensor for highly sensitive detection of bisphenol A *Sens. Actuators B Chem.* **266** 805–11

[46] Zhang K, Song X, Liu M, Chen M, Li J and Han J 2023 Review on the use of magnetic nanoparticles in the detection of environmental pollutants *Water* **15** 3077

[47] Wu Y, Bai H, Zhou Q, Li S, Tong Y, Guo J, Zhou B, Li Z, Zhan Y, Liu M *et al* 2021 Preparation of polyamidoamine dendrimer modified magnetic nanoparticles and its application for reliable measurement of Sudan red contaminants in natural waters at parts-per-billion levels *Front. Chem.* **9** 708995

[48] Sun L, Sun X, Du X, Yue Y, Chen L, Xu H, Zeng Q, Wang H and Ding L 2010 Determination of sulfonamides in soil samples based on alumina-coated magnetite nanoparticles as adsorbents *Anal. Chim. Acta* **665** 185–92

[49] Cao Y, Li J, Luo L, Li P and Liu X 2021 Preparation of magnetic molecularly imprinted nanoparticles for enrichment and separation of 6-benzylaminopurine *Food Res. Dev* **42** 87–93

[50] Kumar A, Joshi H and Kumar A 2022 Arsenate removal from the groundwater employing maghemite nanoparticles *Water* **14** 3617

[51] Jagirani M, Uzcan F and Soylak M 2021 A selective and sensitive procedure for magnetic solid-phase microextraction of lead(II) on magnetic cellulose nanoparticles from environmental samples prior to its flame atomic absorption spectrometric detection *J. Iran. Chem. Soc.* **18** 1005–13

[52] Wu H and Meng L 2018 Liquid chromatography-UV determination of heavy metal ions in environmental samples using dispersive liquid-liquid microextraction coupled with magnetic nanoparticles *Appl. Ecol. Environ. Res.* **17** 1571–84

[53] Tian Y, Xu Z, Yang Y, Wang D, Liu Z and Si X 2022 Magnetic solid phase extraction based on Fe3O4@SiO2@CTS nano adsorbent for the sensitive detection of trace polychlorinated biphenyls in environmental water samples *Microchem. J.* **172** 106947

[54] Ragab G and Bahgat E 2021 Magnetic iron oxide nanoparticles for HPLC determination of valsartan in aqueous environmental samples, biological samples and tablet dosage forms *Int. J. Environ. Anal. Chem.* **101** 2612–28

[55] Onyancha R, Ukhurebor K, Aigbe U, Mogire N, Chanzu I, Kitoto V, Kusuma H and Darmokoesoemo H 2022 A review of the capabilities of carbon dots for the treatment and diagnosis of cancer-related diseases *J. Drug Delivery Sci. Technol.* **78** 103946

[56] Neolaka Y *et al* 2022 Synthesis of zinc (II)-natural zeolite mordenite type as a drug carrier for ibuprofen: drugs release kinetic modelling and cytotoxicity study *Results Chem.* **4** 100578

[57] Eleryan A, Hassaan M, Aigbe U, Ukhurebor K, Onyancha R, Kusuma H, El-Nemr M, Ragab S and El Nemr A 2023 Biochar-C-TETA as a superior adsorbent to acid yellow 17 dye from water: isothermal and kinetic studies *J. Chem. Technol. Biotechnol.* **98** 2415–28

[58] Eleryan A, Aigbe U, Ukhurebor K, Onyancha R, Hassaan M, Elkatory M, Ragab S, Osibote O, Kusuma H and El Nemr A 2023 Adsorption of direct blue 106 dye using zinc oxide nanoparticles prepared via green synthesis technique *Environmental Science and Pollution Research* **30** 69666–82

[59] Kouhestani H and Ebrahimi P 2021 Extraction of carboxin and bensulfuron-methyl using cysteine-functionalized chitosan-coated magnetic nanoparticles and response surface methodology *Sep. Sci. Technol.* **56** 2385–97

[60] García-Merino B, Bringas E and Ortiz I 2022 Robust system for the regenerative capture of aqueous pollutants with continuously synthesized and functionalized magnetic nanoparticles *J. Environ. Chem. Eng.* **10** 108417

[61] Yağci Ö, Akkaya E and Bakirdere S 2020 Nano-sized magnetic Ni particles based dispersive solid-phase extraction of trace Cd before the determination by flame atomic absorption spectrometry with slotted quartz tube: a new, accurate, and sensitive quantification method *Env. Monit. Assess.* **192** 583

[62] He M, Su S, Chen B and Hu B 2020 Simultaneous speciation of inorganic selenium and tellurium in environmental water samples by polyaniline functionalized magnetic solid phase extraction coupled with ICP-MS detection *Talanta* **207** 120314

[63] Leal M, Alonso E, Guerrero M, Cordero M, Pavón J and Torres A 2018 Speciation analysis of inorganic arsenic by magnetic solid phase extraction on-line with inductively coupled mass spectrometry determination *Talanta* **184** 251–9

[64] Turiel E, Díaz-Álvarez M and Martín-Esteban A 2020 Surface modified-magnetic nano-particles by molecular imprinting for the dispersive solid-phase extraction of triazines from environmental waters *J. Sep. Sci.* **43** 3304–14

[65] Cui Z, Xiang L and Tang J 2021 Facile one-pot synthesis of magnetic molecular imprinting polymers as a novel adsorbent for the enrichment of imidacloprid based on a magnetic dispersive micro-solid-phase extraction in water samples *Chem. Pap.* **75** 3787–95

[66] Ragab G and Bahgat E 2021 Magnetic iron oxide nanoparticles for HPLC determination of valsartan in aqueous environmental samples, biological samples and tablet dosage forms *Int. J. Environ. Anal. Chem.* **101** 2612–28

[67] Wu Y, Chen C, Zhou Q, Li Q, Yuan Y, Tong Y, Wang H, Zhou X, Sun Y and Sheng X 2019 Polyamidoamine dendrimer decorated nanoparticles as an adsorbent for magnetic solid-phase extraction of tetrabromobisphenol A and 4-nonylphenol from environmental water samples *J. Colloid Interface Sci.* **539** 361–9

[68] Zheng Y, Liu Y, Guo X, Chen Z, Zhang W, Wang Y, Tang X, Zhang Y and Zhao Y 2020 Sulfur-doped g-C_3N_4/rGO porous nanosheets for highly efficient photocatalytic degradation of refractory contaminants *J. Mater. Sci. Technol.* **41** 117–26

[69] Singh B, Pal Singh C, Panneerselvam P, Jadoun S, Barani M, Ameta S and Ameta R 2022 Synthesis and characterization of Ch-PANI-Fe_2O_3 nanocomposite and its water remediation applications *Water* **14** 3615

[70] Khan S, Irfan S, Lam S, Sun X and Chen S 2022 3D printed nanofiltration membrane technology for waste water distillation *J. Water Process Eng.* **49** 102958

[71] Ali N, Riead M, Bilal M, Yang Y, Khan A, Ali F, Karim S, Zhou G and Wenjie Y 2021 F. Sher and *et al*, 'Adsorptive remediation of environmental pollutants using magnetic hybrid materials as platform adsorbents *Chemosphere* **284** 131279

[72] Ibrahim W, Nodeh H, Aboul-Enein H and Sanagi M 2015 Magnetic solid-phase extraction based on modified ferum oxides for enrichment, preconcentration, and isolation of pesticides and selected pollutants *Crit. Rev. Anal. Chem.* **45** 270–87

[73] Zhao B, He M, Chen B and Hu B 2019 Fe_3O_4 nanoparticles coated with double imprinted polymers for magnetic solid phase extraction of lead(II) *Microchim. Acta* **186** 775

[74] Kong J, Coolahan K and Mugweru A 2013 Manganese based magnetic nanoparticles for heavy metal detection and environmental remediation *Anal. Methods* **5** 5128–33

[75] Maleki B, Baghayeri M, Ghanei-Motlagh M, Zonoz F, Amiri A, Hajizadeh F, Hosseinifar A and Esmaeilnezhad E 2019 Polyamidoamine dendrimer functionalized iron oxide nano-particles for simultaneous electrochemical detection of Pb^{2+} and Cd^{2+} ions in environmental waters *Measurements* **140** 81–8

[76] Wu H, Meng L and Song W 2019 Glutathione-stabilized silver nanoparticles and magnetic nanoparticles combination for determination of lead and cadmium in environmental waters *Sci. Adv. Mater.* **11** 1133–9

[77] Wang X, Yang J, Zhou L, Song G, Lu F, You L and Li J 2021 Rapid and ultrasensitive surface enhanced Raman scattering detection of hexavalent chromium using magnetic Fe_3O_4/ZrO_2/Ag composite microsphere substrates *Colloids Surf. A Physicochem. Eng. Asp.* **610** 125414

[78] Faiz F, Qiao J, Lian H, Mao L and Cui X 2022 A combination approach using two functionalized magnetic nanoparticles for speciation analysis of inorganic arsenic *Talanta* **237** 122939

[79] Xiong Y, Chen S, Ye F, Su L, Zhang C, Shen S and Zhao S 2015 Preparation of magnetic core–shell nanoflower Fe3O4@MnO2 as reusable oxidase mimetics for colorimetric detection of phenol *Anal. Methods* **7** 1300–6

[80] Christus A, Panneerselvam P, Ravikumar A, Morad N and Sivanesan S 2018 Colorimetric determination of Hg (II) sensor based on magnetic nanocomposite (Fe3O4@ZIF-67) acting as peroxidase mimics *J. Photochem. Photobiol. A Chem.* **364** 715–24

[81] Oguz M, Bhatti A and Yilmaz M 2020 Surface coating of magnetite nanoparticles with fluorescence derivative for the detection of mercury in water environments *Mater. Lett.* **267** 127548

[82] Liu Y, Cai Z, Sheng L, Ma M and Wang X 2020 A magnetic relaxation switching and visual dual-mode sensor for selective detection of Hg^{2+} based on aptamers modified Au@Fe3O4 nanoparticles *J. Hazard. Mater.* **388** 121728

[83] Li X, Wen Q, Chen J, Sun W, Zheng Y, Long C and Wang Q 2022 Lanthanide molecular species generated Fe_3O_4@SiO_2-TbDPA nanosphere for the efficient determination of nitrite *Molecules* **27** 4431

[84] Zhang K, Zheng X, Li H and Zhao Z 2020 Human health risk assessment and early warning of heavy metal pollution in soil of a coal chemical plant in Northwest China *Soil. Sediment. Contam. Int. J.* **5** 481–502

[85] Li Q, Dai L, Wang M, Su G, Wang T, Zhao X, Liu X, Xu Y, Meng J and Shi B 2022 Distribution, influence factors, and biotoxicity of environmentally persistent free radical in soil at a typical coking plant *Sci. Total Environ.* **835** 155493

[86] Zhang Y, Guo X, Si X, Yang R, Zhou J and Quan X 2019 Environmentally persistent free radical generation on contaminated soil and their potential biotoxicity to luminous bacteria *Sci. Total Environ.* **687** 348–35

[87] Kim I, Kim J-H, Kim S-M, Park C, Yoon I-H, Yang H-M and Sihn Y 2021 Enhanced selective separation of fine particles from Cs-contaminated soil using magnetic nano-particles *J. Soils Sediments* **21** 346–54

[88] Kasa N, Sel S, Özkan B and Bakırdere S 2019 Determination of palladium in soil samples by slotted quartz tube-flame atomic absorption spectrophotometry after vortex-assisted ligandless preconcentration with magnetic nanoparticle-based dispersive solid-phase micro-extraction *Env. Mont. Assess* **191** 692

[89] Patiño-Ropero M, Díaz-Álvarez M and Martín-Esteban A 2017 Molecularly imprinted core–shell magnetic nanoparticles for selective extraction of triazines in soils *J. Mol. Recognit.* **30** e2593

[90] Kim K, Yoon S, Ahn J, Choi Y, Lee M, Jung J and Park J 2017 Synthesis of fluorescent naphthalimide-functionalized Fe_3O_4 nanoparticles and their application for the selective detection of Zn^{2+} present in contaminated soil *Sensors Actuators* B **243** 1034–41

[91] Yang L, Su P, Chen X, Zhang R and Yang Y 2015 Microwave-assisted synthesis of poly (ionic liquid)-coated magnetic nanoparticles for the extraction of sulfonylurea herbicides from soil for HPLC *Anal. Methods* **7** 3246–52

[92] Pang J, Song X, Huang X and Yuan D 2020 Porous monolith-based magnetism-reinforced in-tube solid phase microextraction of sulfonylurea herbicides in water and soil samples *J. Chromatogr.* A **1613** 460672

[93] Pang J, Chen H and Huang X 2021 Magnetism-assisted in-tube solid phase microextraction for the on-line chromium speciation in environmental water and soil samples *Microchem. J.* **164** 105956

[94] Li W, Wang R and Chen Z 2019 Metal-organic framework-1210(zirconium/cuprum) modified magnetic nanoparticles for solid phase extraction of benzophenones in soil samples *J. Chromatogr. A* **1607** 460403

[95] Hubetska T, Kobylinska N and Menendez J 2020 Application of hydrophobic magnetic nanoparticles as cleanup adsorbents for pesticide residue analysis in fruit, vegetable, and various soil samples *J. Agric. Food Chem.* **68** 13550–61

[96] Kim J, Kim S, Yoon I and Kim I 2020 Application of polyethylenimine-coated magnetic nanocomposites for the selective separation of Cs-enriched clay particles from radioactive soil *RSC Adv.* **10** 21822–9

[97] Hou Q, Ma A, Wang T, Lin J, Wang H, Du B, Zhuang X and Zhuang G 2015 Detection of bioavailable cadmium, lead, and arsenic in polluted soil by tailored multiple *Escherichia coli* whole-cell sensor set *Anal. Bioanal. Chem.* **407** 6865–71

[98] Meng D, Zhao N, Ma M, Fang L, Gu Y, Yao J, Liu J and Liu W 2017 Application of a mobile laser-induced breakdown spectroscopy system to detect heavy metal elements in soil *Appl. Opt.* **56** 5204–10

[99] Jia J, Li H, Zong S, Jiang B, Li G, Ejenavi O, Zhu J and Zhang D 2016 Magnet bioreporter device for ecological toxicity assessment on heavy metal contamination of coal cinder sites *Sensors Actuators* B **222** 90–299

[100] Zhang K, Bao K, Xiong Z and Li T 2022 Didactic experimental design for characterization of biotoxicity in contaminated soils with MNPs-P. phosphoreum biosensor *Exp. Technol. Manag* **39** 209–13

[101] Zhang K, Liu M, Song X and Wang D 2023 Application of luminescent bacteria bioassay in the detection of pollutants in soil *Sustainability* **15** 7351

Chapter 5

Application of magnetic nanomaterials in soil decontamination and bioremediation of contaminants under the influence of a magnetic field

An important ecological challenge that has substantial effects on the environment, human health, and agricultural production is soil contamination, which is driven by urbanization, agricultural and industrial activities, and the inappropriate disposal of waste into the environment. In recent years, there has been concentrated research on the use of nanotechnology (NT) in the remediation of soil contamination using magnetic nanomaterials (MNMs). Various research studies reviewed in this book chapter have shown that these MNMs are effective in the immobilization/sorption of various contaminants found in soil.

5.1 Introduction

Today's agricultural practices have ecological drawbacks, even though they contribute substantially to meeting the constantly increasing worldwide food demand. The expansion of agricultural practices has resulted in widespread soil contamination with a wide range of harmful substances [1]. Fast-growing anthropogenic actions are accumulating potentially toxic metals, agrochemicals, and excess nutrients in the soil. Soil is the foundation of crop production and is enriched with vital macro- and micronutrients that encourage healthy plant growth, ultimately benefiting human well-being. Agronomy nurtures and defines human existence; however, it often disrupts natural ecosystems. The insatiable human hunger for the benefits of natural resources has grown concurrently with the global

doi:10.1088/978-0-7503-6377-8ch5

population increase. There is a conflict between the advantages of agriculture and the sustainable management of agricultural land [2–4].

Agricultural contamination is a broad environmental concern that is generally triggered by the action of many farming practices, such as pesticides and fertilizers, as well as harmful activities and consequences like excessive soil cultivation and runoff. These contaminants arise from the injudicious application of agrochemicals, resulting in a buildup of inorganic contaminants like heavy metals (HMs) and organic contaminants like phenols, halogenated compounds, azo dyes, phthalic esters, insecticides, persistent organic pollutants, and pesticides. These are some of the most persistent problems challenging agricultural production around the globe. These agro-contaminants are extremely toxic and have negative (−ve) effects on air, water, and soil, causing severe toxicity in living organisms and compromising agricultural land fertility via the food chain [1, 5].

An obvious trend arises where higher levels of soil contamination overlap with elevated levels of water contamination, while lower levels of soil contamination align with decreased levels of water contamination. This association emphasizes the probable impact of soil contamination on water quality, highlighting the need for effective remediation procedures to lessen the harmful effects on the environment and human health linked to soil contamination [6].

The optimization of resources and the rehabilitation of damaged soil for agricultural use is essential to strengthen food production to feed 9.9 billion people globally by 2050 [7]. Hence, soil conservation is the greatest responsibility of the modern age due to the pressures of a growing population and the reduction of arable land by anthropogenic actions. Progress in NT has unlocked an international opportunity to remediate or restore contaminated soil in the most efficient way [2].

Creating advanced approaches that can effectively remove agro-contaminants in trace amounts is important. In all fields of science and environmental application, NT methods have been utilized swiftly, with several of these remediation approaches established for both *in situ* and *ex situ* applications due to their unique properties, such as extraordinary surface area (SA), reactive nature, ability to alter physico-chemical properties, and nanoscale dimensions. These nanobioremediation approaches have received greater consideration in recent years owing to their sustainability and effectiveness in curtailing ecological contaminants [1, 5, 8, 9]. The application of nanoparticles (NPs) has been explored in the remediation of pollutants in various ways, such as adsorption, redox precipitation, and coprecipitation due to their large specific SA. With the support of NPs, hyperaccumulators and indigenous soil microbes can enhance the biodegradation process, thus improving the potential level of treatment. These approaches can be described as nanophytoremediation and microbial nanoremediation, respectively [2].

The design and synthesis of nanostructured materials and NPs for ecological pollutant elimination guarantee sustainability, community well-being, and environmental safety [10]. Characteristic magnetic nanomaterials (NMs) or magnetic NPs (iron (Fe), nickel (Ni), cobalt (Co), manganese (Mn), and their corresponding oxides), with their great magnetic properties, are currently applied for the remediation of contaminated soil. They are known for their ease of separation, good biocompatibility, and high sorption capacity. These MNPs can also be combined with materials like metal/metal oxide (MeO) NPs, polymers, graphene, carbon nanotubes, and chitosan through surface functionalization to create nanocomposites or hybrid NPs used for the decontamination of soil in the agronomic sector [11, 12].

The objective of this chapter is to assess and merge the current body of information on the application of magnetic nanoparticles (MNPs) for the treatment of contaminated soil under the influence of an applied magnetic field (AMF). This chapter will further discuss the various sources and types of soil contaminants, NT, and different NM applications, as well as the nanoremediation of contaminated soil and the mechanisms involved.

5.2 Sources and types of soil contamination

Fertile soils are essential for the biosphere to sustain agricultural processes that provide adequate food [13]. Soil is a complex and dynamic environment. It is one of the Earth's main carbon (C) reservoirs, containing both organic and inorganic materials. It delivers an extensive array of important environmental services, including resources (timber, food, fiber, and freshwater), cultural benefits (spiritual and visual values), support (plants, animals, and physical support of human infrastructure), and regulation (prevention of erosion, climate control, and flood mitigation) [1]. Carbon-based soil contains the decayed remains of animal biomass and plants, which are concentrated in the dark topmost layer. The chemical and physical weathering of the parent rock materials over thousands of years results in the creation of inert soils consisting of rocky components. Soil contamination is triggered by the presence of synthetic chemicals or natural variations of soil in the ecosystem [13, 14].

The current conversion of productive land into infertile space by unrestrained urbanization, road and house construction, and industrial substances results in negative ($-$ve) soil contamination. Tertiary contamination or landscape contamination arises from household and construction waste discharged into the soil by various settlements, thereby making these settlements unpleasant and isolating neighboring land. The direct input of pesticides, fumigants, and herbicides used during planting, the application of chemical fertilizers, and atmospheric air contaminants washed out by rain result in positive or point source contamination of soil [15, 16].

Contaminated soils represent a secondary source of pollutants discharged to the neighboring ground and surface water, the air, and eventually oceans [16]. Soil contamination is caused by the presence of anthropogenic chemicals or the natural modification of soil within the ecosystem. Normally, this arises from ruptured underground storage links, pesticide application, polluted surface water percolating down to subsurface strata, or fuel and oil dumping. Waste leachates are released into the soil as a result of the degree of industrialization and the intense usage of chemicals such as pesticides, solvents, heavy metals (HMs), petroleum hydrocarbons, etc. As reported by Havugimana *et al* [13], natural contamination of soil is caused by natural phenomena such as landslides, hurricanes, floods, and earthquakes, resulting in natural harm to the soil composition. The most predominant pollutants found in the soil are heavy metals (HMs), which are generally generated by industrial and other associated activities [13, 17].

Pollutants in the soil arrive at their end points in different ways. Most carbon-based pollutants can undergo chemical transformation or be degraded into products that may be more or less complex than the original compound. It should be noted that chemical elements like HMs cannot be broken down, but their properties may be transformed; hence, they can more or less easily be taken up by animals or plants. Various pollutants differ in their propensity to end up in water held in the soil or the underlying groundwater by discharge via the soil, to be volatilized into the air, or to be firmly bound to the soil. The extent of pollutants is also impacted by the soil properties and whether the pollutants are readily taken up by animals or plants. Site management and the use of land for horticultural practices can impact some of the characteristics of the soil. Key soil properties like soil texture and pH, the quality of the organic matter in the soil, the moisture level, temperature, and other chemicals present may impact pollutant behavior [13]. The environmental characteristics of soils, such as their structure, skeletal nature, humus content, reactivity, edaphon, presence of external materials, and the availability of nutrients, substantially affect their production, filtering, buffering, and other functions [1]. Considering these properties is vital for evaluating susceptibility and forecasting potential hazards. A significant role is also played by weather patterns, such as wind direction and precipitation [6].

According to Havugimana et al [13], the important ideas in understanding soil pollution and its impacts on ecological health depend on the soil horizons. This comprises the soil's development from the parent material (bedrock) as a result of mechanical weathering of rock by abrasion, moving water, glaciers, temperature variation, lichens, wind, and chemical weathering events. Under perfect climatic conditions, soft parent materials are transformed into 1 cm of soil within 15 years. Additionally, within the O and A horizons, a large number of fungi, earthworms, small insects, and bacteria form a multifaceted food network in the soil, recycling soil nutrients and contributing to soil fertility. In the B horizon (subsoil), there is less carbon-based material and fewer organisms than in the A horizon. The C horizon consists of fragmented bedrock, which does not contain any carbon-based materials. The soil pH, as well as the rate of water absorption and retention are determined and influenced by the chemical composition. Finally, in the R horizon, the layer of unweathered rock (bedrock) is located below all the other layers [13].

5.3 Soil contamination remediation approaches

Remediation of soil is a procedure that is employed to eliminate and reduce pollutants to restore soil for plants and healthy environments. To remediate soil, it is essential to comprehend its nature, biological diversity, and organic matter. Hence, it is possible to remediate soil contamination using a sequence of various technologies such as chemistry, microbiology, biology, geology, environmental science, and engineering, which are viable and efficient [18]. Soil pollution remediation is a key task in the preservation of the natural environment and human health. To address this problem, several approaches have been applied to remediate the contamination of soil with various pollutants. The methods employed in the remediation of contaminated soil or its restoration comprise chemical, physical, thermal, mixed, and biological approaches. By their mode of action, they are categorized as containment, transport, and transformation. The physical approaches include procedures like soil washing, thermal treatment, excavation, and screening. Chemical approaches involve adsorption on surfaces, chemical extraction from the soil, oxidation/reduction, and neutralization. Biological approaches like bioaugmentation and phytoremediation employ microorganisms and plants to reduce and absorb contaminants from the soil [18, 19].

The selected soil remediation method is achieved by confining the polluted site, capping polluted soil, excavating the soil from the polluted site, and treating soil contaminants from the polluted location. The remediation approaches on sites are based on *in situ* and *ex situ* processes of remediation, as displayed in figure 5.1. In the *in situ* remediation approach, the soil being remediated remains in its original location during remediation. This involves the addition of an agent to facilitate the treatment of the polluted soil. The key benefit of the *in situ* treatment category is that it allows for the soil to be remediated without altering its original environmental factors like organic matter, biodiversity, and structure. This approach encompasses stabilizing, biodegrading, or separating contaminants from soil using microbes, soil microorganisms, and plants. Table 5.1 shows the benefits and drawbacks of using the *in situ* approach for soil treatment [19, 20]. The excavation and transportation of polluted soil to off-site treatment facilities are not necessary in the *in situ* approach; hence, there is negligible soil disturbance and reduced exposure of workers and the surrounding public to the pollutants, and the treatment costs may be substantially diminished. However, certain field conditions like soil permeability, depth of pollution, potentially unknown leaching of chemicals, and weather must be carefully considered [18].

The *ex situ* approach involves more conventional treatment methods. In the *ex situ* methods, technologies require the polluted soil to be excavated and transported for remediation. In these approaches, landfarming and biopiles may include bioremediation in the treatment procedure. The choice of an *ex situ* bioremediation approach is generally based on the cost of treatment, the depth of pollution, the type of toxin, the pollution concentration, the topographic area, and the characteristics of the contaminated site [19, 20].

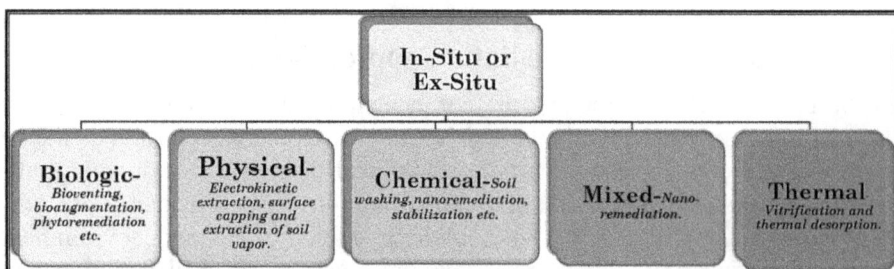

Figure 5.1. Various *in situ/ex situ* approaches used for soil remediation.

Table 5.1. Different *in situ* soil treatment approaches, including some of their benefits and drawbacks.

In situ approach	Principle	Advantages/drawbacks
Biological approaches		
Monitoring natural attenuation	Reduction of pollutants by biological, physical, and chemical processes; observation of the effectiveness of soil cleanup and hazard reduction	Cost-efficient but lengthy procedure; only appropriate for some pollutants
Biostimulation	Produces a favorable setting for microbial pollutant degradation; involves the addition of electron acceptor/donor, nutrients, oxygen; requires temperature and pH control.	Cost-efficient; only applicable for certain pollutants; low pollutant bioavailability decreases degradation effectiveness
Bioaugmentation	Addition of supplemented/adapted microbes that can degrade a specific pollutant	Can be unproductive owing to the failure to adapt microbes to ecological settings
Phytoremediation	Stockpiling of pollutants in underground plant parts; pollutants detached in crops or respired into the air via transpiration	Green, low cost, and non-hazardous; pollutants captured remain in the soil
Bioventing (biosparging)	Injection of air/oxygen above or below the water table at low pressure; venting rate is improved to maximize biodegradation	Negligible disturbance of site; inappropriate in low-porosity soils or mixed soils.
Nanobioremediation	Bioremediation enabled by NMs improves microorganisms' growth, modulates the remediating agents, and induces creation of remediating microbial enzymes.	Quicker than bioremediation alone; cost-efficient; appropriate for several pollutants; distribution of NMs to reach the pollutants is challenging; NMs may have toxic impacts on organisms
Chemical approaches		
Chemical reduction/ oxidation	Oxidizing or reducing agents are injected into the soil to cause comprehensive or incomplete degradation of pollutants	Swift degradation of high pollutant concentrations; non-selective degradation can cause negative ecological impacts
Physical approaches		
Extraction of soil vapor	Application of high vacuum to the soil induces airflow, eliminates volatile pollutants, and enhances aerobic settings	Cost-efficient approach; limited effectiveness in damp and impenetrable soil

Electro-treatment	Electric field applied to the soil causes movement and the desorption of charged pollutants; may improve the transport of nutrients, degrade microbes, and enhance bioavailability	Most effective in fine-grained and wet soils and also appropriate for inert pollutants

The choice of an applicable approach for contaminated soil remediation depends on different factors, including the type and concentration of contaminants present in the soil, the nature and source of the soil, the mass or level of pollution, the type of soil, the depth and extent of pollution, and fiscal and time limitations. Contaminated soil remediation can offer substantial benefits, such as decreasing human health risks, preserving the ecosystem, enhancing soil quality for agriculture and other applications, and increasing the value of land [6, 21]. However, the astronomical cost and extensive time required by these procedures pose substantial challenges in this sector. In addition, the selection of an appropriate approach for specialized environments requires thorough consideration and assessment of the environment. Each individual approach is strengthened by its benefits, removal effectiveness, targeted pollutants, and constraints [6, 18].

5.4 Nanoremedation of contaminated soil using MNPs

The contemporaneous challenges of meeting the growing food production requirements are critical to prevent further soil degradation from strongly impacting agricultural yields. Nanoenabled soil treatments have emerged as auspicious and justifiable solutions for rejuvenating compromised soil resources. NT is a developing paradigm in agriculture specifically aimed at improving plant phytoremediation capabilities for soil, demonstrating its potential in the agronomic area. NPs offer diverse benefits over conventional soil treatment approaches, mainly owing to their size and surface area [1].

MNPs, combined with their adapted behavior, are distinctively linked with the nanoscale. Their improved reactivity arises from their enormous surface-area-to-volume ratio and their intrinsic magnetic properties such as superparamagnetism (SPM). This compelling combination enables MNPs to act as effective sorbents for contaminants and allows them to be magnetically separated from the surrounding environment for further processing and removal. The intrinsic reactivity of iron (Fe) may be utilized in contaminated soil treatment by reducing the contaminants to less toxic compounds. The ecological use of MNPs for the elimination of pollutants can take place via chemical reactions, physical adsorption, or detection via chemo-sensing [21].

Yet, employing NT in soil treatment requires certain modifications. First, engineered NMs should be capable of being transported to the polluted soil areas. Furthermore, the transported NPs should be confined within the contaminated area where they capture the pollutants. Nevertheless, there is a possibility for the NPs to aggregate into micrometer to millimeter masses. This reduces their large specific SA and ease of transport to the polluted zones. Thus, polymers like carboxymethyl cellulose, starch, and other collagens are often conjugated on the surfaces of NPs to improve their stability and prevent them from aggregating. This is due to the steric hindrance affecting the polymer molecules, which prevents particle aggregation. Coating and modifying MNPs forms a core–shell structure that permits the removal of specific metallic and chemical pollutants [17, 21–23].

Pollutants like HMs in the soil infiltrate the soil layers and react with other soil elements via sorption and desorption. Thus, it is extremely desirable to hinder the movement of pollutants into the soil via immobilization approaches. NMs can effectively capture pollutants from water and soil due to their unique properties like reactivity, elevated SA, and exceptional ability to bind with other materials. Applying NT in the remediation of soil and water involves certain modifications. Essentially, engineered NMs should be able to be transported to polluted soil or water sites. Furthermore, the transported NPs should be limited within the polluted precinct where they capture the pollutants. But, there is a greater prospect for the NPs to amass into micrometre to millimetre masses, thereby reducing their features like their detailed SA and ease of transferability to the polluted region. Hence, polymers like carboxymethyl cellulose, starch, and other collagens mostly bind to the surface of the NPs to improve their stability and prevent them from amassing [17].

MNMs play a substantial role in processes associated with the fertility of the soil, the restoration of soil, and the ultimate growth of plants. They have been assessed as additives for improving the fertility of the soil and for the magneto-assisted elimination of noxious pollutants and other detrimental organic compounds, thereby promoting soil restoration [24]. The effective immobilization of contaminants like HMs in the soil is impacted by the different physicochemical properties of soil, including soil pH, soil aggregates, cation exchange capacity (CEC), and organic matter content. The impact of various NMs and contaminants like HMs on soil physicochemical properties varies, and this should be considered when applying NMs for the treatment of soil contaminated by HMs. Depending on the ecological pollutant's physicochemical and structural properties, it can be sorbed by surfaces as a result of intermolecular forces such as electrostatic interaction/attraction, chemisorption (covalent bonding), and van der Waals forces [25, 26].

In soil contaminated with arsenic (As) and polycyclic aromatic hydrocarbons (PAHs), Fe_3O_4 NPs were employed to immobilize these contaminants from the soil. Based on the Toxicity Characteristic Leaching Procedure (TCLP) defined in the US Environmental Protection Agency (EPA) Technique 1311, there were 43% and 92% reductions of As immobilized using 1% and 5% of Fe_3O_4 NPs, respectively. As depicted in figure 5.2, as adapted and reproduced from Baragaño et al [27], a decrease was also noted in the content of total petroleum hydrocarbons (TPHs 49%) and PAHs (89%), which was observed at low amounts of NPs used to immobilize them. This study also reported that after the immobilization of the contaminated soil by the NPs, there was no substantial impact on soil properties like pH and electrical conductivity. Based on the phytotoxicity of NPs in the soil after treatment, it was found that the presence of Fe in the soil did not substantially impact the germination test [27].

In a study by Ghasemi et al [28], Fe_3O_4 NPs modified with ethylene diamine tetraacetic acid (EDTA) were explored for the removal of HMs (Ag^{1+}, Hg^{2+}, Mn^{2+},

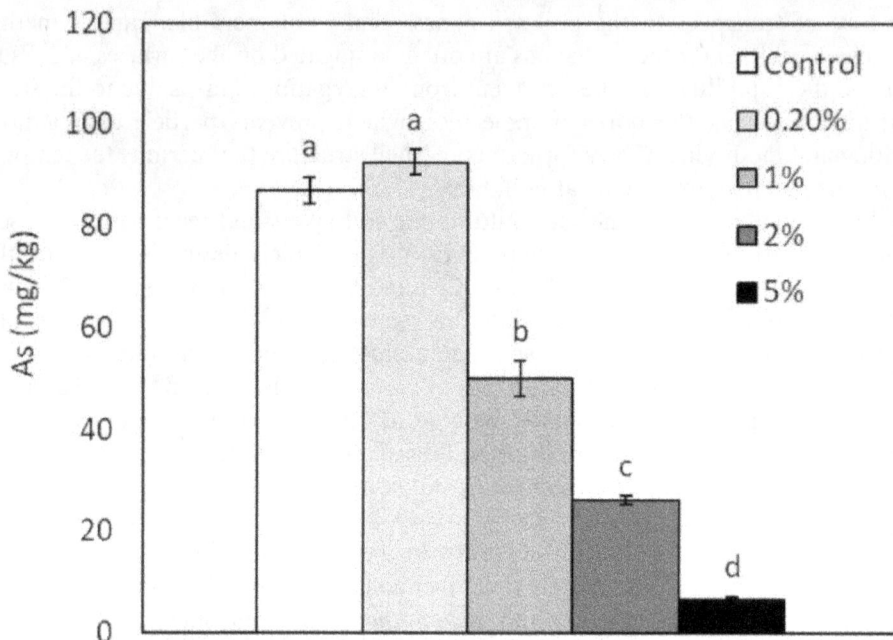

Figure 5.2. Leachability of As at various doses of Fe_3O_4 NPs in the TCLP test. Reprinted from [27], Copyright (2020), with permission from Elsevier.

Zn^{2+}, Pb^{2+}, and Cd^{2+}) from ecological soil samples using the Box–Behnken (BB) experimental strategy. It was noted in this study that 99% of these HM ions were removed from the soil sample within 10 min; hence, the prepared nanocomposites were proposed as an effective nanosorbent for the decontamination of soil containing these HMs. Based on a study by Baragaño et al [29], Fe-based NMs (goethite nanospheres, nGothe) and zero-valent Fe NPs (zVFe NPs) were assessed for the immobilization of As in polluted soil using 0.5%–10% doses of the nGothe and zVFe NPs. An estimated 90% of As was reported to have been removed from the polluted soil at a dosage of 2% of zVFe NPs, with no undesirable impacts on the soil properties. It was also reported that the soil phytotoxicity decreased, but a slight increase in As^{3+} was noted at elevated doses of material. Using 0.2% of nGothe, a slight reduction in the percentage of As removed from the soil (82%) was noted, with increased phytotoxicity at elevated doses. This was attributed to the noticeable improvement in the electrical conductivity. In a study by Peng et al, zVFe NPs, ferriferous oxide (Fe_3O_4 NPs), and ferrous sulfide (FeS NPs) were explored, in addition to biochar (BC)-modified zVFe NPs (zVFe@BC), BC-modified Fe_3O_4 NPs (Fe_3O_4@BC), and BC-modified FeS NPs (FeS@BC) for the removal of lead (Pb) from soil and to examine the resulting impact on the soil microecosystem. The percentage concentrations of Pb ions immobilized from the soil at day 90 using zVFe NPs, Fe_3O_4 NPs, FeS NPs, zVFe@BC, Fe_3O_4@BC, and FeS@BC were approximately 46%, 48%, 3%, 55%, 5%, and 31%, respectively. It was also observed that the pH value of the soil was substantially reduced by the use of FeS NPs and FeS@BC but increased due to the use of the other Fe NPs. There was an improvement in soil enzyme activities due to the use of the Fe_3O_4 NPs and Fe_3O_4@BC groups, but most of these activities were inhibited in the other Fe-based NP groups, with the determined bacterial number assessed using the FeS NPs group [30].

Fe NPs prepared using the leaf extract of *Euphorbia cochinchinensis* were used to reduce the amounts of As^{3+} and As^{5+} in polluted soil. After 120 days of inoculation, there was a reduction in the nonspecifically bound and specifically bound As elements by 27% and 67% due to the use of phytogenic Fe NPs. The application of the phytogenic Fe NPs reduced the amounts of As^{3+} and As^{5+}, with dominant species of As^{5+} remaining in the soil [31]. The efficiency of various Fe NPs, such as zVFe NPs, bimetallic zVFe–palladium (Pd), and Fe_3O_4 NPs, was assessed for the removal of Cr and polychlorinated biphenyls (PCBs) from soil. A substantial (98%) decrease in the concentration of Cr in water-soluble extracts was reported for the Fe NPs after three sampling periods compared to the control samples. A decreased Cr leachability in the sample of soil treated and the immobilization of Cr on zVFe NPs and zVFe–Pd resulted in the sorption of Cr^{6+} on the shell of the materials, and Cr^{3+} reduction was reported in this study. There was a substantial reduction of approximately 68% of the PCBs in the soil samples following treatment with the three classes of NPs used in this study after 15 days [32]. Fe_3O_4 NPs were explored for the immobilization of the radionuclide ^{137}Cs in contaminated clay taken from the soil. This study found that an increase in the Fe_3O_4 NP dosage resulted in an increase of more than 90% in clay immobilization at acidic pH values. At low pH values, the positive charges on the Fe_3O_4 NPs increased, leading to the binding of the negatively charged clay particles due to electrostatic interaction/attraction. The application of sieving magnetic separation under acidic conditions was shown to enable the efficient removal of about 60% of the Cs-contaminated clay from the soil [33].

Fe_2O_4 NPs stabilized with manganese (Mn) were explored for the immobilization of As^{3+} in soil. The Fe_2O_4@Mn NPs achieved a substantial increase in As^{3+} immobilization, showing an 88% reduction in plant bioavailability, a 64% decrease in physiologically based extraction, and 90% retention under TCLP testing according to the soil amendment procedure test results. From the sequential extraction approach, the As bound to Fe_3O_4@Mn was found to increase from 70% to 82% upon applying 1%–3% of Fe_3O_4@Mn [34]. Fe_3O_4 NPs synthesized via the coprecipitation approach were also applied in the removal of Cr, Fe, Ni, Zn, Cu, and Pb from soil and water samples. These metals were optimally immobilized in the soil at pH 0.7 for 20 min, with removal efficiencies varying between about 70%–100% [35]. Fe_3O_4 NPs stabilized on polyacrylic acid were also explored for the removal of Cd^{2+} from polluted soil in a study by Mohamadiun *et al.* The unsurpassed removal of 100% Cd^{2+} ions from polluted soil was achieved at 500 ppm of nanofluid concentration, a pH of 6.5, a contact time of 24 h, and a polluted soil mass (g) ratio to nanofluid volume (ml) ratio of 1:150 [36]. Employing the green synthesis approach, Fe_3O_4 NPs were prepared using the extract of *Eucalyptus globulus* as an organic reducing agent. The prepared FeO NPs were assessed for the reduction of various HM levels (Cr^{6+} and Cd^{2+}) in polluted soil. A 100% removal of Cr^{6+} and Cd^{2+} ions from the polluted soil by the FeO NPs was reported in this study after 30 min at pH values of 2–6. This resulted in the reduction of Cr^{6+} to Cr^{3+} following the application of the NPs to the polluted soil. This was ascribed to the transfer of electrons, leading to the oxidation of Fe_3O_4 NPs to Fe^{2+} and Fe^{3+} [37].

5.5 Influence of environmental factors on the removal of contaminants from soil using MNPs

The surrounding ecosystem has a large effect on the remediation of contaminants in soil and water using NMs. The mechanism of reducing waste in the ecosystem is controlled by parameters like pH, contact time, and temperature, as shown in figure 5.3, adapted and reproduced from Sathish *et al* [17]. A key parameter in the eradication of pollutants such as HMs from the ecosystem using NMs is temperature. With an increase in temperature, there is an increase in the rate of the redox reaction, thereby resulting in a decreased rate of adsorption [17]. A governing parameter in the sorption of contaminants by NPs from the ecosystem is the peripheral pH. The effectiveness of the sequestration of pollutants like HMs from water and soil is generally impacted by variations in pH. At neutral pH settings, precipitation and adsorption rates are enhanced. Individual NMs vary in their behavior even at similar pH values. Another factor that regulates the elimination of contaminants from polluted sites is the contact time between the pollutants and the NMs. With increasing contact time, the sorption rate of pollutants by NMs

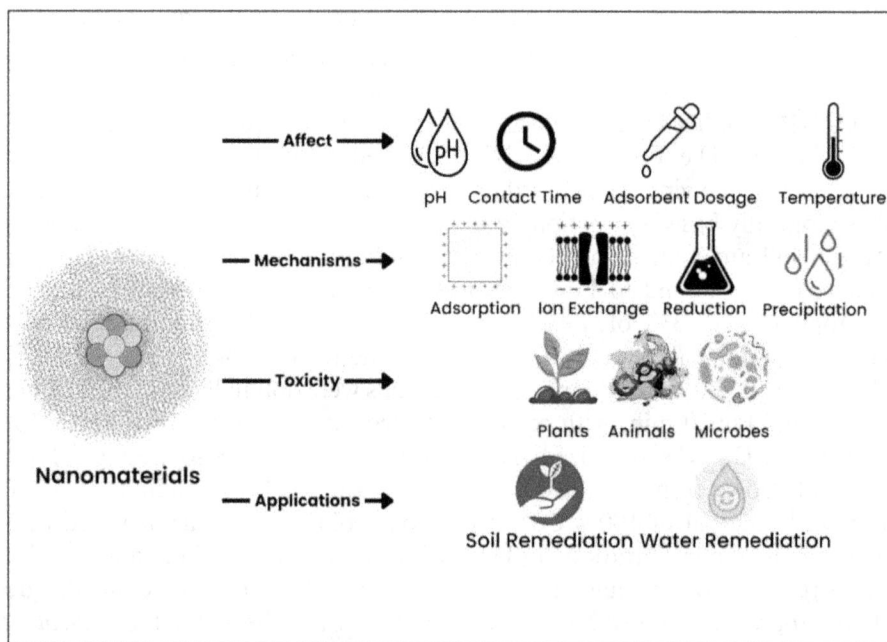

Figure 5.3. Factors impacting the remediation of polluted soil and water using NPs. Reproduced from [17]. CC BY 4.0.

improves. Kinetic models are generally employed to evaluate the impact of contact time. However, the growth trend changes to equilibrium after a threshold contact time is reached [17, 38].

5.6 Conclusions

In recent times, there has been an increased discharge of contaminants into the environment. The continued release of these contaminants poses a widespread threat to the environment and human health. MNMs are important and capable remediation materials for the treatment of various soil pollutants. MNPs are effective when applied in soil treatment, as their outstanding sorption capacity, combined with their extraordinary magnetic susceptibility, provides a perfect mechanism for their magnetic recovery. This chapter has focused on the effective application of MNPs for the remediation/immobilization of contaminants in soil due to their extraordinary specific SA and reactivity. Several research studies in this chapter have illustrated that these MNPs are effective in the immobilization/sorption of various contaminants, especially HMs found in soil.

References

[1] Dhanapal A, Thiruvengadam M, Vairavanathan J, Venkidasamy B, Easwaran M and Ghorbanpour M 2024 Nanotechnology approaches for the remediation of agricultural polluted soils *ACS Omega* **9** 13522–33

[2] Rajput V, Minkina T, Upadhyay S, Kumari A, Ranjan A, Mandzhieva S, Sushkova S, Singh R and Verma K 2022 Nanotechnology in the restoration of polluted soil *Nanomaterials* **12** 769

[3] Rajput V *et al* 2022 A review on nanobioremediation approaches for restoration of contaminated soil *Eur. J. Soil Sci.* **11** 43–60

[4] Ubani O, Modise S and Atagana H 2023 Magnetic nanomaterials for decontamination of soil *Magnetic Nanomaterials: Synthesis, Characterization and Applications* (Cham: Springer International Publishing) pp 171–86

[5] Hidangmayum A, Debnath A, Guru A, Singh B, Upadhyay S and Dwivedi P 2023 Mechanistic and recent updates in nano-bioremediation for developing green technology to alleviate agricultural contaminants *Int. J. Environ. Sci. Technol.* **20** 11693–718

[6] Dehkordi M, Nodeh Z, Dehkordi K, Khorjestan R and Haffarzadeh M 2024 Soil, air, and water pollution from mining and industrial activities: sources of pollution, environmental impacts, and prevention and control methods *Res. Eng.* **23** 102729

[7] Medina-Pérez G, Fernández-Luqueño F, Vazquez-Nuñez E, López-Valdez F, Prieto-Mendez J, Madariaga-Navarrete A and Miranda-Arámbula M 2019 Remediating polluted soils using nanotechnologies: environmental benefits and risks *Polish J. Environ. Stud.* **28** 1013–30

[8] Aigbe U, Ukhurebor K, Onyancha R, Okundaye B, Aigbe E, Kusuma H, Noto L, Osibote O and Atagana H 2023 Applications of magnetic nanomaterials for wastewater treatment *Magnetic Nanomaterials: Synthesis, Characterization and Applications* (Cham: Springer International Publishing) pp 129–69

[9] Aigbe U and Osibote O 2020 A review of hexavalent chromium removal from aqueous solutions by sorption technique using nanomaterials *J. Environ. Chem. Eng.* **8** 104503

[10] Targuma S, Njobeh P and Ndungu P 2021 Current applications of magnetic nanomaterials for extraction of mycotoxins, pesticides, and pharmaceuticals in food commodities *Molecules* **26** 4284

[11] Le Wee J, Law M, Chan Y, Choy S and Tiong A 2022 The potential of Fe-based magnetic nanomaterials for the agriculture sector *ChemistrySelect* **7** e202104603

[12] Woldeamanuel M, Mohapatra S, Senapati S, Bastia T, Panda A and Rath P 2024 Role of magnetic nanomaterials in environmental remediation *Iron Oxide-Based Nanocomposites and Nanoenzymes: Fundamentals and Applications* (Cham: Springer International Publishing) pp 185–208

[13] Havugimana E, Bhople B, Kumar A, Byiringiro E, Mugabo J and Kumar A 2017 Soil pollution–major sources and types of soil pollutants *Environ. Sci. Eng.* **11** 53–86

[14] Batool M, Cihacek L and Alghamdi R 2024 Soil inorganic carbon formation and the sequestration of secondary carbonates in global carbon pools: a review *Soil Syst.* **8** 15

[15] Titirmare N, Gaikwad A and Margal P 2023 Soil pollution and enviornmental health *Advances in Soil Science* (New Delhi: Bright Sky Publications TM) 267–307

[16] 2021 Sources of Soil Pollution *Global assessment of soil pollution: Report* (Rome: Food and Agriculture Organization of the United Nations) 3

[17] Sathish T, Ahalya N, Thirunavukkarasu M, Senthil T, Hussain Z, Siddiqui M, Panchal H and Sadasivuni K 2024 A comprehensive review on the novel approaches using nano-materials for the remediation of soil and water pollution *Alex. Eng. J.* **86** 373–85

[18] Ghule M and Ramteke P 2022 Soil chemical pollution and remediation *Hazardous and Trace Materials in Soil and Plants* (New York: Academic) 57–71

[19] Hussain A, Rehman F, Rafeeq H, Waqas M, Asghar A, Afsheen N, Rahdar A, Bilal M and Iqbal H 2022 In-situ, ex-situ, and nano-remediation strategies to treat polluted soil, water, and air—a review *Chemosphere* **289** 133252

[20] Romantschuk M, Lahti-Leikas K, Kontro M, Galitskaya P, Talvenmäki H, Simpanen S, Allen J and Sinkkonen A 2023 Bioremediation of contaminated soil and groundwater by *in situ* biostimulation *Front. Microbiol.* **14** 1258148

[21] Pratt A 2014 Environmental applications of magnetic nanoparticles *Frontiers of Nanoscience* **6** (Amsterdam: Elsevier) 7 259–307

[22] Das P, Mohanty C, Purohit G, Mishra S and Palo S 2022 Nanoparticle assisted environ-mental remediation: applications, toxicological implications and recommendations for a sustainable environment *Environ. Nanotechnol., Monitor. Management* **18** 100679

[23] Karn B, Kuiken T and Otto M 2009 Nanotechnology and *in situ* remediation: a review of the benefits and potential risks *Environ. Health Perspect.* **117** 1813–31

[24] Spanos A, Athanasiou K, Ioannou A, Fotopoulos V and Krasia-Christoforou T Functionalized magnetic nanomaterials in agricultural applications *Nanomaterials* **11** 3106

[25] Zhang Y, Zhang Y and Wu A 2024 Design and construction of magnetic nanomaterials and their remediation mechanisms for heavy metal contaminated soil *Sci. Total Environ.* **951** 175369

[26] Bradford S, Shen C, Kim H, Letcher R, Rinklebe J, Ok Y and Ma L 2022 Environmental applications and risks of nanomaterials: an introduction to CREST publications during 2018–2021 *Crit. Rev. Environ. Sci. Technol.* **52** 3753–62

[27] Baragaño D, Alonso J, Gallego J, Lobo M and Gil-Díaz M 2020 Magnetite nanoparticles for the remediation of soils co-contaminated with As and PAHs *Chem. Eng. J.* **399** 125809

[28] Ghasemi E, Heydari A and Sillanpää M 2017 Superparamagnetic $Fe_3O_4@$ EDTA nanoparticles as an efficient adsorbent for simultaneous removal of Ag (I), Hg (II), Mn (II), Zn (II), Pb (II) and Cd (II) from water and soil environmental samples *Microchem. J.* **131** 51–6

[29] Baragaño D, Alonso J, Gallego J, Lobo M and Gil-Díaz M 2020 Zero valent iron and goethite nanoparticles as new promising remediation techniques for As-polluted soils *Chemosphere* **238** 124624

[30] Peng D, Wu B, Tan H, Hou S, Liu M, Tang H, Yu J and Xu H 2019 Effect of multiple iron-based nanoparticles on availability of lead and iron, and micro-ecology in lead contaminated soil *Chemosphere* **228** 44–53

[31] Su B, Lin J, Owens G and Chen Z 2020 Impact of green synthesized iron oxide nanoparticles on the distribution and transformation of As species in contaminated soil *Environ. Pollut.* **258** 113668

[32] Gil-Díaz M, Pérez R, Alonso J, Miguel E, Diez-Pascual S and Lobo M 2022 Iron nanoparticles to recover a co-contaminated soil with Cr and PCBs *Sci. Rep.* **12** 3541

[33] Kim I K J, Kim S, Park C, Yoon I, Yang H and Sihn Y 2021 Enhanced selective separation of fine particles from Cs-contaminated soil using magnetic nanoparticles *J. Soils Sediments* **21** 346–54

[34] Zialame A, Jamshidi-Zanjani A and Darban A 2021 Stabilized magnetite nanoparticles for the remediation of arsenic contaminated soil *J. Environ. Chem. Eng.* **9** 104821

[35] Singh A, Chaudhary S and Dehiya B 2021 Fast removal of heavy metals from water and soil samples using magnetic Fe3O4 nanoparticles *Environ. Sci. Pollut. Res.* **28** 3942–52

[36] Mohamadiun M, Dahrazma B, Saghravani S and Darban A 2018 Removal of cadmium from contaminated soil using iron (III) oxide nanoparticles stabilized with polyacrylic acid *J. Environ. Eng. Landsc. Manag.* **26** 98–106

[37] Andrade-Zavaleta K, Chacon-Laiza Y, Asmat-Campos D and Raquel-Checca N 2022 Green synthesis of superparamagnetic iron oxide nanoparticles with *Eucalyptus globulus* extract and their application in the removal of heavy metals from agricultural soil *Molecules* **27** 1367

[38] Jiang D, Zeng G, Huang D, Chen M, Zhang C, Huang C and Wan J 2018 Remediation of contaminated soils by enhanced nanoscale zero valent iron *Environ. Res.* **163** 217–27

Chapter 6

Effect of magnetic fields in gas and contaminant sensing using magnetic nanomaterials

To maintain high environmental standards, management, monitoring, and sustainability practices are required. A crucial aspect of this is controlling potentially hazardous substances in the air, water, and soil. In order to reduce harmful substances in the environment, magnetic nanomaterials (MNMs) have emerged as a viable alternative due to their low toxic effects, good electrical conductivity, ease of functionalization, and ease of production. Hence, the prospect of using magnetic gas sensors (MGSs) in particular is reviewed in this chapter. Future considerations and insights about MGSs vis-à-vis MNMs for environmental management and the monitoring of hazardous contaminants are also highlighted.

6.1 Introduction

Humanity has always been interested in natural resources, especially those related to water (given that water is one of the most essential natural resources), but these have recently become a serious problem in virtually every aspect of human life [1–5]. Domestic water, groundwater, surface-level water (lakes, rivers, rain, springs), and seawater (oceans, seas) are the four primary categories of resources related to water [3]. The need for drinking water and other water resources for industrial, biomedical, and agricultural purposes has grown dramatically with worldwide population growth, industrialization, and civilization trends. The pollution of surface water is the result of various harmful contaminants, such as organic pollutants, heavy metals (HMs), medical or pharmaceutical contamination, dyes, etc. being released into water resources [6–13].

The creation of different pollutants and industrial effluents is a major problem nowadays, since it contaminates both groundwater and surface water resources [3, 14–16]. HMs and other pollutants can occasionally be found in high concentrations in industrial waste materials; these contaminants need to be removed to reduce their detrimental impact on living organisms, as well as the environment. During the water management and treatment process, their acidic nature is frequently neutralized, HMs are eliminated or diminished, and other contaminants are lowered to a level that is acceptable. According to several studies, the following categories of harmful chemicals are most frequently found in water resources: biological (bacteria, viruses, and fungi), chemical (HMs, aromatic molecules, toxic organic aliphatics, dyes, pesticides, detergents, surfactants, insecticides, nitrogen (N) and sulfur (S) compounds, etc.), and physical (nanoplastics, microplastics, etc.) [3, 17–24].

In order to protect human welfare through food safety and minimize severe dangers, it is crucial to regulate dangerous contaminants in water resources and enhance the continuous surveillance of toxic constituent contaminants in water-related environments and products [25–29] [25–29]. Water resources can be monitored, analyzed, and treated using a variety of techniques and technologies [30–34]. In practice, technologically advanced and dependable analytical methods and instruments, such as electrochemical biosensors (BSs), solid-phase extraction (SPE), spectrophotometry, gas chromatography, high-performance liquid chromatography, and mass spectrometry, depend on the physical and chemical characteristics of the analytes [35–37].

While these methods have numerous benefits, they also have numerous limitations, including the requirement for a skilled operator, expensive maintenance, and labor-intensive sample preparation. Strict sample preservation guidelines must be followed in order to maintain the specimens in their initial state and avoid the creation of by-products. Furthermore, the concentration of volatile substances may differ from that of the reference analytical samples. It goes without saying that volatile substances and solvents might react, causing unanticipated consequences [3]. Online monitoring is hardly possible because results take a long time to produce [36]. Due to their quick and sensitive analysis of various water pollutants, electrochemical BSs mitigate these limitations when compared to other techniques. However, electrochemical BSs do have limitations, such as environmental sensitivity and difficulty in detecting small concentrations of pollutants due to weak oxidation–reduction signals. The goal of the modified BSs embellished with magnetic nanoparticles (MNMs) is to improve response due to the presence of iron ions. When it comes to capturing the desired molecules within complicated analyte samples that comprise a multitude of undesirable particles in water-based environments such as rivers, drinking water, seas, sewage, etc. MNMs are superior to nonmagnetic substances [38–40].

Water treatment approaches such as electric flotation, filtration, gravity separation, biological treatment, chemical coagulation, chemical precipitation, reverse osmosis, activated carbon adsorption, air flotation membrane bioreactors, photocatalysis, and oxidation, as well as the use of nanoparticles (NPs)/nanomaterials (NMs), as opposed to nanotechnology (NT)/bionanotechnology (BNT), currently constitute the predominant methods used in water treatment [23, 35, 41–45]. Every technique and technology has advantages and disadvantages, though. For instance, although chemical precipitation, coagulation, and flocculation are economical methods, they produce toxic waste. More importantly, using these methods to remove waste necessitates an extra step that raises operating costs. Compared to alternative methods, the employment of NMs in various architectures for the treatment of drinking water is comparatively quick and economical and offers a number of benefits. This presents opportunities to cut expenses, time, and labor while effectively removing contaminants and recycling industrial waste [42, 43, 46, 47].

Recent developments in NT/BNT, especially in the areas of NMs and membranes/nanofiltration, have shown, for example, how beneficial detectors, disinfectants, and purification devices are in terms of efficiency, affordability, speed, and ease of manipulation when used for recycling and treating wastewater-related materials [3, 48]. These will ultimately benefit both developed and developing nations greatly. In light of their many advantages over other methods for detecting water resources, including their large number of active sites, adsorption by hydroxyl groups on the NM surface, outstanding biocompatible nature, rapid operation, easy preparation, magnetic separation of liquid/solid separation, reuse/recycling, and capacity for separating/recognizing particular molecules and cells, MNMs have garnered a lot of interest recently [3, 49–52].

The characteristics of MNMs could potentially be applied in organic and mineral pollutants, photocatalysts [6, 22, 53–58], magnetic BSs for monitoring, detecting, tracking, and measuring pollutants in water-based environments, as well as other environmental settings [3–5, 59]. Furthermore, these materials' non-toxic properties and environmentally favorable characteristics open up the prospect of using them to detect dangerous chemicals in water. A few of the numerous areas in which magnetic composites find use are targeted drug delivery to tumors in medical treatment, enhancing contrast in MRI scans, pigments and dyes for the paint industry, information storage devices for electronic audio/video equipment, radar-absorbing materials, electromagnetic technologies, etc [3, 51]. By generating surface bonding in their outermost shell, oxidation–reduction processes, or hydroxide deposition on MNMs (which promotes the bonding of diverse functional groups), MNMs can efficiently manage pollutants. MNMs can also be used for monitoring and eliminating HM cations such as lead (Pb), cadmium (Cd), zinc (Zn), nickel (Ni), copper (Cu), chromium (Cr), arsenic (As), and cobalt (Co); organic chemicals such as nitro-aromatic composites, chlorinated dyes, and pesticides; and anionic composites, for example, chromates, nitrates, and phosphates [22, 52, 60–62]. Consequently, this chapter provides an overview of some MNMs that can be utilized for detecting, monitoring, and identifying pollutants, specifically in MGSs. The monitoring and detection of environmental contaminants using MNMs as nanosensors (NSs) or nanobiosensors (NBSs) will be addressed in the next section of this chapter. In conclusion, research developments and future perspectives involving MNMs are highlighted in relation to environmental applications and the detection of potentially hazardous substances in water resources.

6.2 An overview of concepts, techniques, and advancements in the field of magnetic gas sensors

Since the development of the first sensor by Clark and Lyons in 1962 [63], several designed and produced sensors, NSs, or NBSs have been reported in the last few years [64–67]. Several benefits and applications for sensors, NSs, or NBSs can be found in a range of industries, including agriculture [68–70], biomedicine [71–74], and the environment [75–78]. Like every other sensor, NS, or NBS, gas sensors (GSs) function by converting the surface-induced influences of gas adsorption into a signal that can be sensed, detected, and monitored based on the changes/modifications in the material's electrical, optical, mechanical, thermal, magnetic (including spin and magnetization), and piezoelectric properties [59]. Several procedures, including the Hall effect, magnetization, spin orientation, ferromagnetic resonance, magneto-optical Kerr effect, and magnetostatic wave or magnetostatic spin-wave (MSW) oscillation effect, are utilized in MGS to measure the modifications in the magnetic features of the active and dynamic materials [59]. The limitations of various gas sensing types include their susceptibility to dampness, high operating temperatures, and chemical selectivity. When it comes to chemiresistive-type GSs, for instance, the presence of different gas species might cause the sensor resistance to alter significantly in the real world, leading to a complicated overall electrical impact because of concurrent surface reactions [59]. Furthermore, dependable detectors for powdered samples are difficult to build using traditional electrical-property-based GSs. Given that electrical-property-based hydrogen GSs are combustible at higher operating temperatures, fire safety is another concern. In order to address these problems, concepts for MGS have been developed. In these concepts, the presence of gas molecules modifies the magnetic characteristics of the materials [59].

For MGS applications, a variety of magnetic effects are used, including the magnetoplasmonic effect, the MSW effect, ferromagnetic resonance (FMR), magnetization, Hall, Kerr, and spin change effects, as presented in figure 6.1, which shows the magnetic special effects employed for the fabrication and production of gas sensing devices (as adapted and reproduced from Shinde and Rout [59]), namely: (a) the Hall effect, as reported by Ciprian *et al* [79]; (b) the Kerr effect, as reported by Ciprian *et al* [79]; (c) the ferromagnetic resonance (FMR) effect, as reported by Lueng *et al* [80]; (d) the magnetoplasmonic effect, as reported by Ignatyeva *et al* [81]; (e) the magnetic moment or spin effect, as reported by Hsu *et al* [82]; and (f) the MSW effect, as reported by Matatagui *et al* [83]. For MGS applications, several magnetic materials have been used, such as powders, thin films, and NMs with magnetic (ferro- and antiferromagnets), diluted magnetic semiconductors (DMS), and Pd alloys with transition metals (Co, Ni, Fe, Mn, or Cu). Owing to the following reasons, MGS have become an increasingly attractive choice compared to their counterparts based on electrical properties [59, 84–87]. MGSs can be employed at any temperature by selecting magnetic materials with different Curie temperatures (Tc). Since no electrical interactions are required for detection, the magnetic response is significantly quicker than that of chemiresistive sensors. Regarding the gas to be sensed, there is a reduced possibility of an explosion due to ignition if MGSs are utilized in hydrogen-powered vehicles or in the presence of reactive composites or contaminants. The magnetic response is significantly quicker than that of chemiresistive sensors.

Notwithstanding, this chapter will not go into details about the concepts, techniques, and advancements in MGSs but will only attempt to highlight the

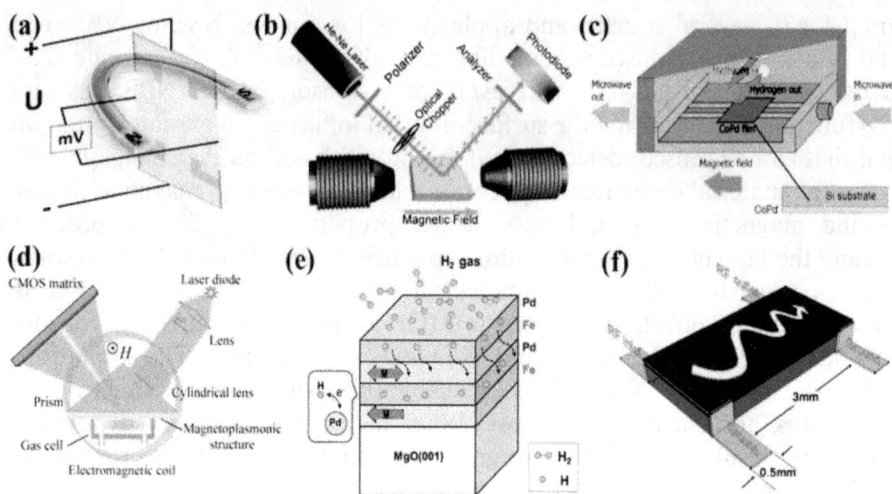

Figure 6.1. Magnetic special effects utilized for the fabrication and production of gas sensing devices. (a) Hall effect. Reproduced from [59]. CC BY 4.0. (b) Kerr effect. Reproduced from [79] with permission from the Royal Society of Chemistry. (c) FMR effect. Reprinted from [80], Copyright (2019), with permission from Elsevier. (d) Magnetoplasmonic effect. Reproduced from [81]. CC BY 4.0. (e) Magnetic moment or spin effect. Reproduced from [82]. CC BY 4.0. (f) MSW effect. Reproduced from [83] with permission from the Royal Society of Chemistry.

applications of MGSs for environmental management and the monitoring of hazardous contaminants. Hence, see Shinde and Rout [59] for more details of the concepts, techniques, and advancements in MGS, where they reviewed the fundamental concepts, detection processes, and the most recent advancements in magnetic gas sensing. They gave an in-depth description of the various effects used in MGS applications, such as the magnetoplasmonic effect, MSW effect, FMR effect, magnetization, Hall effect, Kerr effect, and spin change effects. Additionally, they highlighted the recent advancements in the field of magnetic sensing related to various materials, such as Pd alloys containing transition metals such as Co, Ni, Fe, Mn, or Cu; powders; thin films; and NMs with magnetic (ferro- or antiferromagnetic) properties. They also covered the history of sensitivity increases as well as the operation of MGSs.

6.3 The applications of magnetic gas sensors for environmental management, monitoring, detection, and identification of hazardous contaminants

6.3.1 Magnetic nanobiosensors for monitoring, detection, and identification of various contaminants in water-related resources

One of the largest issues facing the world today is the disposal of contaminants in water-related resources, which poses a major risk to the existence of life. Owing to the processes of the industrial revolution, civilization, and growing urbanization, precise and comprehensive information regarding the movement, occurrence, and deterioration, as well as the measurement, detection, and identification of water contaminants, is important. Researching, managing, and surveying these hazardous contaminants has attracted a lot of attention. Since the concentrations of various pollutants change throughout different water resources, multipollutant identification and assessment need to be done simultaneously and with great sensitivity, reliability, and innovation. In practical applications, analytical pathways and devices such as electrochemical BSs, SPE, spectrophotometry, gas chromatography, high-perform-ance liquid chromatography, and mass spectrometry depend on the physical and chemical characteristics of the analytes [37, 88–90]. These methods have several benefits, but they also have limitations. For instance, several of these approaches necessitate the use of an experienced operator; others have maintenance costs that are prohibitively expensive; and yet others involve difficult sample preparation. Furthermore, samples may be inaccessible due to a lack of precision or experience. In order to maintain the specimens in their original condition and avoid the formation of by-products, sample maintenance is necessary. The standard labora-tory concentration of the samples may differ from the measured amount of volatile chemicals. The sample needs to dissolve in the specified solvents, and the volatile substances may react with the solvents, generating undesirable consequences. Monitoring via the Internet may not be possible because results take a long time to produce [36]. Due to their quick and responsive assessment of various contam-inants in water, electrochemical BSs mitigate these limitations when compared to other techniques. However, electrochemical BSs/NSs have a few limitations as well, such as their small oxidation–reduction signals when measuring environmental contaminants, which lower detection capability. As a result, customized sensors embellished with MNMs are being developed to exhibit enhanced response due to the presence of iron ions. When it comes to capturing target molecules in complicated analyte specimens that comprise a multitude of undesirable particles (such as sewage, drinking water, oceans, or rivers), applying MNMs has an advantage over nonmagnetic substances. It is difficult to retrieve nonmagnetic materials, and this creates issues in the subsequent analysis if they are employed after detecting, monitoring, identifying, measuring, and absorbing pollutants that affect water.

Figure 6.2 illustrates how MNMs/magnetic composites (magnetic molecular imprinted polymers, MMIPs) are separated after being applied to complex analytes in order to detect, monitor, identify, measure, and absorb different pollutants using a magnetic field (MF), as adapted and reproduced from Aylaz *et al* [91]. In recent years, magnetic NBSs have drawn a lot of interest [39, 92], especially in water-related resource management, detection, and identification [3].

6.3.2 Monitoring, detection, and identification of pharmaceutical pollutants

The potentially hazardous direct release of pharmaceuticals into water-related resources reduces water quality and renders it unfit for consumption [1, 3, 7, 93]. Additional health problems caused by these pharmaceutical-based chemicals include typhoid fever, cholera, polio, diarrhea, and dysentery [3, 94]. A method for quickly

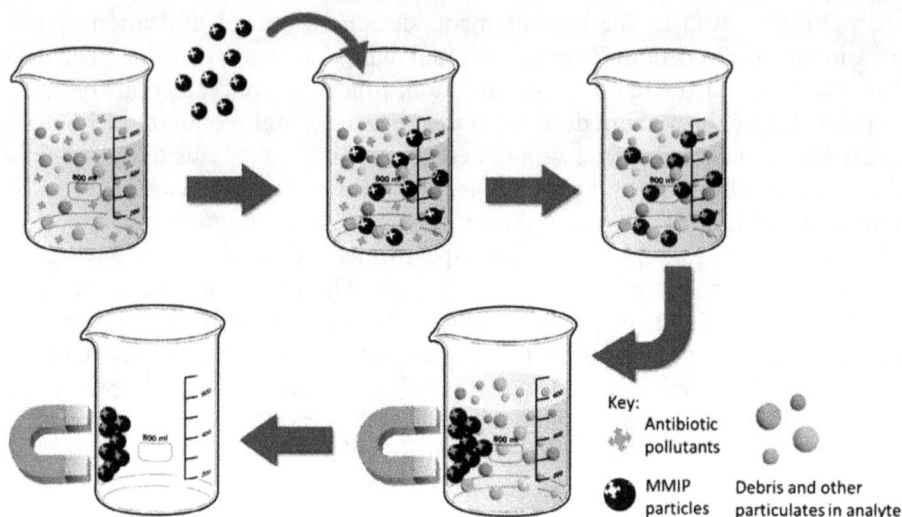

Figure 6.2. The representation of MMIPs/MNMs are separated by an MF after their utilization in monitoring, detecting, identifying, measuring, and adsorbing different pollutants from complex analytes [91]. John Wiley & Sons. [© 2021 Society of Chemical Industry].

detecting a small amount of *Escherichia coli* (*E. coli*) in water for consumption using bacteriophage (BP)-conjugated magnetic beads (MBs) was demonstrated by Chen *et al* [95]. This method for detecting *E. coli* in water for consumption is schematically depicted in figure 6.3, as adapted and reproduced from Chen *et al* [95]. Following their separation and concentration, the BPs invaded and lysed the bacterial host cells, releasing β-galactosidase. Colorimetric techniques made it simple to identify the released enzyme. After only 2.5 h, this method could monitor and detect or sense *E. coli* at a concentration of 104.00 colony-forming units (CFUs) ml^{-1}.

Considering the benefits of using a BP, which include ease of use and propagation, affordability, a wide host range, and outstanding protection for both human and animal forms, this approach is evidently more advantageous than antibodies, since it has the ability to identify every instance of *E. coli* found in water-related resources [95]. A different study [96] demonstrated the capability to use aptamer 'bacteria-sized' gold (Au)-coated magnetic microdisks for removing *E. coli* from samples of potable water at levels of 102.00 CFU 100 ml^{-1} in under forty-five minutes. Jia *et al* [97] produced a magnetic aptasensor that used changes in the time relaxation of superparamagnetic iron oxide NMs to identify *Pseudomonas aeruginosa* (*P. aeruginosa*) quickly. A recovery test was used for monitoring and evaluating the various concentrations of *P. aeruginosa* in potable water in order to determine the viability and dependability of a sensitive magnetic aptasensor. As demonstrated by a recovery rate of up to 108.90% in the results obtained, the aptasensor was able to accurately determine and identify *P. aeruginosa* in potable water. MNMs have been reported to be useful in other investigations for the detection of pharmaceutical pollutants in water-related resources [98–101]. In order to detect oxytetracycline (OTC) in seawater, Zhao *et al* [102] developed a highly sensitive NBS using a freshly

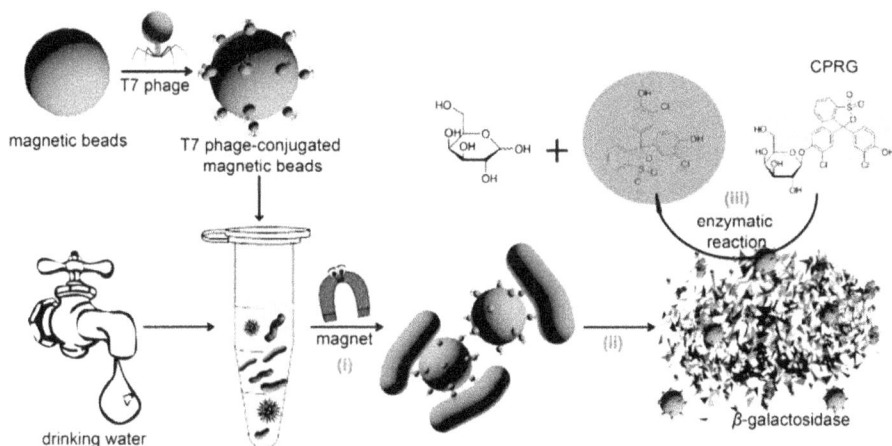

Figure 6.3. Representation of the detection of *E. coli* in potable water using BP-conjugated MBs. (i) Removal of *E. coli* from potable water using BP-conjugated MBs; (ii) infection of *E. coli* by BPs, which results in the discharge of β-galactosidase; (iii) to read out colorimetric information, the hydrolysis of chlorophenol red-β-D-galactopyranoside is catalyzed by β-galactosidase. Reprinted with permission from [95]. Copyright (2015) American Chemical Society.

imprinted molecular magnetic polymer. After determining the amount of OTC in seawater that was collected from the Jiaozhou Bay Gulf, these researchers used multivariate optimization to screen important variables and then produced a core–shell NM structure that comprised Fe_3O_4 NMs and a polymer. The findings indicated a suitably linear response dependence on spike concentration ranging from approximately $3.00–100.00 \, \mu g \, l^{-1}$, indicating that none of the specimens contained any OTC. The recovery rate ranged from 89.75% to 107.65%, with an average standard deviation of less than 5.44% for the seawater samples. The limit of detection (LOD) was determined to be $0.70 \, \mu g \, l^{-1}$.

A class of chemical drugs known as sulfonamides has both antibacterial and antimicrobial resistance potential. Disposing of these medications into soil or water-related resources is too risky because they have the potential to cause allergies and cancer as chemotherapeutic agents. In order to produce an exceptionally selective core–shell magnetic (Fe_3O_4-chitosan) molecularly surface-imprinted polymer capable of tracking and breaking down a combination of antibiotics composed of sulfonamide groups that are intricate template particles; specifically, sulfamethoxazole (SMZ) and sulfamethazine (SFZ) in drinking water and wastewater (sewage) samples, Qin *et al* [103] used surface-imprinted technological advancements. Their findings proved that there had been no detection of sulfonamides (at detectable amounts) in drinking water. SMZ concentrations of 112.00 and $64 \, ng \, l^{-1}$ were observed in sewage samples. It was determined that this compound is more effective at SMZ detection in water-related resources. In a different study, Zamora-Gálvez *et al* [104] described the use of MNMs adorned with imprinted molecules to identify a particular sulfonamide in seawater samples taken from the Poblenou Beach in the Mediterranean Sea without requiring a label. Electrochemical impedance spectroscopy, which makes use of electrodes printed on a screen to measure a material's impedance, revealed an incredibly sensitive monitoring range of 1.00×10^{-12} mol l^{-1} (about 2.80×10^{-4} ppb), which was comparable to results obtained using mass spectrometry and liquid chromatography. Additionally, it was mentioned that due to the magnetic qualities of the MNMs, this method facilitated preconcentration, separation, and analyte modification and had outstanding sensitivity, making it suitable for monitoring seawater. More details of selected pharmaceutical pollution monitoring, detection, and identification studies using MNMs in water-related resources are contained in [3] by Hojjati-Najafabadi et al.

6.3.3 Monitoring, detection, and identification of pesticide pollutants

The need for food is increasing as a result of growing populations. To meet this demand, agricultural businesses are exposed to pesticides, hormone products, herbicides, chemical-based fertilizers, and other unpredictable hazards to the environment. While pesticides fall under the category of organic pollutants, most scientific publications have examined their significance and effects on the environment separately [105–107]. Dichlorodiphenyltrichloroethane (DDT), for instance, is categorized as both a persistent organic pollutant and a pesticide. Pesticides are substances that destroy pests and can take any material form. One of the most pressing environmental problems is the contamination of water-related resources by pesticides, as these substances tend to enter the water at minimal levels and act slowly [108].

While the impacts of these synthetic substances are long-lasting and life-threatening, pesticides are difficult to identify and assess in the environment. One class of organic chemicals used as a substitute for chlorine-containing hydrocarbons is organophosphorus (ONP) insecticides. With regard to the risks to human well-being, environmentalists are increasingly concerned about water contaminants containing ONP compounds [109–111]. Xie et al [111] developed a novel hybrid NM known as $Fe_3O_4@mSiO_2$ microspheres. Surface water was sourced from several locations in Shanghai, and ONP compounds were found in drinking water, tap water, and surface water. These findings demonstrated that surface water was not being checked for malathion, parathion-methyl, or diazinon. A further research study was carried out by Nasiri et al [112], where they prepared and produced a magnetic graphene-based NM coated with polyvinyl alcohol (PVA) and employed it for analyzing the residues of ONP composites in samples of water taken from an area in Iran. They found that outstanding extraction effectiveness, as well as a small LOD of around 20.00–80.00 pg ml^{-1}, was attained because of the hybrid NM's high surface area and magnetic characteristics. Furthermore, the NM may be useful for extracting and preconcentrating several species, including hormonal substances, herbicides, insecticides, and amino acids. The newly developed magnetic graphene-based hybrid NMs were synthesized by Rashidi Nodeh et al [110] to monitor both polar and nonpolar organophosphate pesticides simultaneously. Their findings demonstrated that the LOD was well below the recommended residual level (100 pg ml^{-1}) for organophosphate pesticides in water samples set by the European

Union. The linear ranges for diazinon and chlorpyrifos were found to be 10–100 pg ml^{-1}, while those for dimethoate and phosphamidon were found to be 100–1000 pg ml^{-1}. The absorption capacity was also computed using the Langmuir equation. In addition, a hypothesized absorption mechanism was proposed. The suggested process explains how the graphene-based hybrid NMs offer robust π-π stacking and sensitivity to nonpolar ONP pesticides with benzene rings in the structures of diazinon and chlorpyrifos. Additionally, the hybrid NM uses H-bonding and electrostatic attraction to affect the polar dimethoate and phosphamidon. Ben Attig et al [113] detected additional pesticides, specifically N-methylcarbamate, in surface water samples in a sensitive and selective manner. They devised an analytical approach that involved detecting a capillary diode electrophoresis array for the purpose of selectively identifying the pesticides, followed by combining a temperature-controlled ionic liquid with magnetic multiwalled carbon nanotubes (MWCNTs). By using the ionic liquid–magnetic MWCNT process, this technique reduced the amount of volatile/unstable organic solvents required and the time needed to prepare the sample without centrifugation. In addition, for MNMs, the adsorption period was decreased due to electrostatic attraction. High temperatures and complex mechanisms were also unnecessary. Phase separation was accomplished with ease through the use of an external MF. Sample handling and isolation/ offline detection were this route's primary disadvantages and constraints. Under ideal circumstances, the required limits were attained within the 5.60–7.30 ng ml^{-1} range, and the degree of recovery ranged from 85.00% to 102.40%. The authors suggested developing and refining this high-sensitivity detection technique to track more chemicals used in agriculture [113].

Organochlorine pesticides are extensively used in agriculture all over the world and have been produced in mass quantities for many years in an effort to control malaria and typhoid fever [114–116]. Zhou et al [117] produced polyamidoamine magnetic dendrimers using a high-performance liquid chromatography method to determine and magnetically separate organic chlorinated pesticides from selected Chinese parks in an environmentally friendly way. The findings gathered showed that the water samples had no organic chlorinated pesticide residues. The aforementioned approach exhibited outstanding accuracy and effectiveness, and the anticipated recoveries ranged from 90.6% to 103.5%. This approach may replace methods used for evaluating other pesticides. Additionally, the separation was accelerated by the application of an external MF, which reduced the need for lengthy analysis. The use of organic solvents makes this extraction approach less expensive than other traditional techniques such as liquid–liquid extraction. More details of studies of the monitoring, detection, and identification of pesticide pollutants using MNMs in water-related resources are contained in a review by Hojjati-Najafabadi et al [3].

6.3.4 Monitoring, detection, and identification of organic contaminants

Organic contaminants in water-related resources have the potential to adversely affect people, wildlife, and the environment. The primary component of dangerous organic pollutants is a substance that has been present in the environment for an extended period and contains elements like carbon, hydrogen, nitrogen, etc. Such pollutants can quickly contaminate water supplies and lead to long-term health issues for people. Polluted water supplies often include volatile chemicals, nitrogen-containing compounds, phenolic composites, and phthalates [23, 118–121]. Wang *et al* [121] produced a magnetic hybrid nanoporous 3D graphene 3DG-ZnFe$_2$O$_4$ NM. Bisphenol F (BPF), bisphenol S (BPS), bisphenol C (BPC), tetrabromobisphenol A (TBBPA), tetrachlorobisphenol A (TCBPA), bisphenol A (BPA), bisphenol AP (BPAP), and bisphenol AF (BPAF) were among the nine bisphenol analogues that were found in the water sample using the NM. The results indicated that the 3DG-ZnFe$_2$O$_4$ NM had an outstanding influence on detecting organic contaminants in the water samples; under ideal circumstances, the recovery rate ranged from 95.10% to 103.80%, and the LOD was approximately 0.05–0.18 ng ml^{-1}. They discovered that the NM increased the effectiveness of magnetic extraction and the detection of nine bisphenols. In order to determine the levels of organic micropollutants (OMPs) such as organochlorine pesticides (OCPs), polycyclic aromatic hydrocarbons (PAHs), nitrobenzenes (NBs), and phthalate esters (PAEs) in the water of the Huai River in China, Liang *et al* [32] developed magnetic microspheres (M150) as an SPE method. Their results were then compared to those obtained using hydrophilic-lipophilic balance (HLB) and an octadecyl-type hydrocarbon with 18 carbon atoms (C18). According to the findings, M150 had greater extraction ability and recovery effectiveness than HLB and C18 for a specific set of OMPs, which included NBs, PAEs, PAHs, and OCPs. The effectiveness of recovery increased when the amount of MNMs was increased to 80 mg ml^{-1}, the pH was 2, and the treatment time was 60 min. All of the contaminants discovered had an analytical LOD of M150 of below 3.90 ng l^{-1}, which is regarded as a low level. Thus, the findings presented indicate the excellent precision and viability of this approach.

Using a similar approach to this technique for detecting organic contaminants, Xu *et al* [122] applied a magnetic microporous organic framework (an NM) to identify phenyl urea herbicides in water samples taken from Baoding, Hebei, China. Their approach utilized high-performance liquid chromatography (HPLC) with UV for detection. The data gathered indicated that 0.03 ng ml^{-1} of monuron, 0.10 ng ml^{-1} of chlorotoluron, and 0.10 ng ml^{-1} of metoxuron were found in lake water. The innovative NM constituted an efficient means of detecting OMPs in water-related resources. Zhao *et al* [123] used a magnetic imprinted nanoscale polymer together with UV–vis analysis to accomplish a spectrophotometric assessment of malachite green in sediments from a marine environment. They came to the conclusion that the analytical characteristics were reliable and the LOD was low, validating the wide linear range of accepted techniques for detecting malachite green in sediment samples from the marine environment. The influence of microsphere NMs (Fe$_3$O$_4$/GO/Ag) for the detection and breakdown of methylene blue (MB) in water samples was quantitatively analyzed by He *et al* [124] using surface-enhanced Raman scattering (SERS). They subsequently reached the conclusion that the Fe$_3$O$_4$/GO/Ag microspheres demonstrated exceptional and quantitative SERS sensitivity to track aromatic particles, with a LOD of less than 10^{-9} M and an outstanding linear relationship, demonstrated by an MB correlation coefficient (R^2) of approximately 0.91 for SERS intensities. In the future, the developed Fe_3O_4/GO/ Ag microspheres may potentially prove to be an important tool for efficiently monitoring and detecting organic contaminants. Further details of some organic pollutant monitoring, detection, and identification studies on the use of MNMs in water-related resources are contained in a review by Hojjati-Najafabadi *et al* [3].

6.3.5 Monitoring, detection, and identification of heavy metals

While very small amounts of HM ions, such as Pb, As, Hg, Cd, Cr, etc. are required and beneficial for the maintenance of good health in the biological components of human and animal bodies, these HM ions tend to bind with N, S, and oxygen (O) [3]. As a result, HMs have emerged as one of the major categories of dangerous water contaminants that have the potential to endanger public health and cause environmental issues. Controlling and monitoring these kinds of water quality issues is, therefore, very important. MNs commonly used to manage and detect HM ions in water resources are iron-oxide-based MNMs derived from Fe^{2+} and Fe^{3+} species [3]. In MNMs, iron cations induce electron hopping and superexchange reactions that boost the conductivity of the materials, particularly Fe_2O_3. This can enhance the monitoring effectiveness of magnetic NBSs. When used in NSs, MNMs can aggregate, reducing their ability to detect water contaminants [125, 126]. To overcome this limitation, different material species have been used to functionalize the MNMs [127– 129]. In this context, Chimezie *et al* [130] used differential pulse anodic stripping voltammetry (DPASV) to monitor, detect, and identify As^{3+} in drinking/domestic water, lake water, and mineral water, utilizing a magnetic reduced graphene-oxide NM as an electrochemical NS/BNS. They came to the conclusion that As^{3+} could be monitored in water-related resources using the electrochemical NBS, as the sensitivity and LOD were, under ideal circumstances, 0.33 and 0.10 $\mu g\,l^{-1}$, respectively. Some scientific papers have reported the use of MNMs in the monitoring, detection, and identification of HMs, including Cd and Pb cations [128, 129, 131, 132]. Mohammadi *et al* [132] gathered three natural water samples in the Iranian province of Kerman: drinking/domestic water, well water, and industrial wastewater. Then, using an electrode modified with a carbon nanotube–MNM (Fe_3O_4)–eggshell composite carbon paste, the concentrations of Cd ions in the water samples were determined by DPASV. Their study showed that the recommended electrode could preferentially monitor the Cd ions in different real samples, with a linear response range of 0.50–250 $ng\,l^{-1}$ and an observed/detected LOD of 2.40 $ng\,l^{-1}$ under ideal circumstances. The simultaneous monitoring, detection, and identification of HM pollution in water-related resources has drawn a lot of attention, since it saves time as well as being economical. In this context, Baghayeri *et al* [131] developed and produced a new glassy carbon electrode (GCE) functionalized graphene-oxide-based magnetic electrochemical NBS to simultaneously monitor, detect, and identify Pb^{2+} and Cd^{2+} in water-related resources, namely Chalidarreh Lake and Kardeh River, both in Iran. They studied and optimized the effects of several experimental factors on carbon electrodes that were magnetically altered. The improved electrochemical NBS showed a detection correlation coefficient (R^2) of more than 0.99 under ideal circumstances, resulting in linear voltammetry response ranges of 0.20–140.00 $mg\,l^{-1}$ for Cd ions and 0.40–120.00 $mg\,l^{-1}$ for Pb ions. The LODs for Pb(II) and Cd(II) were found to be 130.00 and 70.00 $ng\,l^{-1}$, respectively. Additionally, they demonstrated that significant interference was not caused by a number of metal cations, such as Cu^{2+}, Tl^{1+}, Hg^{2+}, Zn^{2+}, Mg^{2+}, Co^{2+}, Ni^{2+}, Mn^{2+}, Ca^{2+}, Cr^{3+}, In^{3+}, and Fe^{3+}. The sensitivity of the electrochemical NBS is exceptional, enabling the monitoring, detection, and identification of specific analyte ions.

In a different study, Wang et al [133] used differential pulse voltammetry (DPV) to monitor, detect, and identify Cu^{2+} and Cd^{2+} in natural water simultaneously. Their electrochemical NBS was based on thiol and a magnetic polymer, and it showed suitable suppression of interference capability along with excellent sensitivity owing to the mutually beneficial effect of a magnetic component integrated into the polymer. The developed NBS can function in a variety of real-world work environments owing to the suppression of interference produced by coexisting cations and anions. Conversely, a highly sensitive electrochemical NBS for detecting, monitoring, and identifying HM ions was designed using polydopamine as a covalent modification. Electrochemical reduction of the collected HM ions to zero-valent metals occurred at steady potentials of roughly 0.20 V for Cu^{2+} and 0.60 V for Cd^{2+}. Under optimum conditions, the linear response and LOD for Cu^{2+} and Cd^{2+} within their respective concentration ranges were found to be 0.2 and 0.1 and 0.5–50 $\mu g\,l^{-1}$, respectively. Fan et al [134] developed an electrochemical NBS, a GCE modified with a high-surface-area magnetic spinel NM ($ZnFe_2O_4$ (ZFO)), for the simultaneous measurement and monitoring of Cu^{2+}, Pb^{2+}, and Hg^{2+} in natural water resources. The ZFO–GCE electrode demonstrated exceptional sensitivity as well as effectiveness with these ions, resulting in a linear response and LODs of 12.03, 7.38, and 1.61 nM for Cu^{2+}, Pb^{2+}, and Hg^{2+}, respectively. Outstanding suppression of interference analysis and stability, as well as consistency and reproducibility, were exhibited by the ZFO–GCE electrode, which was effectively used in water-related resource domains. The ZFO–GCE electrodes, they concluded, might be commonly employed for the simultaneous monitoring, measurement, detection, and identification of several HM ions; in this respect, they agreed with several other authors [135–137]. A highlight of some research into MNMs for the detection, monitoring, and identification of HM ions in water-related resources is provided by Hojjati-Najafabadi et al [3].

6.4 Conclusions and future insights

The disposal of potentially harmful substances in water-related resources is steadily rising as a result of industry, civilization, and urbanization and is now a major global issue everywhere, particularly in water-rich nations. It puts aquatic organisms and human well-being in imminent danger. Consequently, it is imperative to obtain precise and comprehensive data regarding the movement, occurrence, and deterioration, as well as the identification, quantification, monitoring, detection, and measurement of water contaminants. Because of recent developments in NT/

BNT, the application of NMs to manage water contaminants in water-related resources is yielding numerous benefits and financial gains for both developed and developing nations.

Due to their characteristics, including low levels of toxicity, good electrical conductivity, ease of functionalization, large numbers of active sites, adsorption by hydroxyl groups on the NM surface, as well as their outstanding biocompatibility, quick processing, ease of preparation, magnetic liquid/solid separation, reuse and recycling, and capacity for separating and recognizing specific composites and cells, MNMs have garnered a lot of interest as a potential substitute for monitoring hazardous pollutants over the past few years. However, employing MNMs as NBSs to monitor water contaminants is not without its difficulties. The agglomeration of MNMs, the memory effect in signal recording, and the detection of NM contaminants at low concentrations present the most significant hurdles for the use of magnetic NBSs.

This chapter has shed light on the toxicity of dangerous water-related organic contaminants, pharmaceuticals, pesticides, and HMs in water-related resources, as well as magnetic NBSs for contamination control and monitoring. As described in detail, MNMs offer a number of benefits related to toxic effects, functionalization, biological compatibility, processing, preparation, separation, cost, and their ability to identify and monitor contamination of water sources. The efficacy of MNMs as NBSs is diminished by the possibility of MNM agglomeration. The functionalization of MNMs with different material species effectively resolves this issue. In summary, MNMs that were included in various matrices performed admirably as NBSs to measure and identify diverse types of contamination in surfaces, soil, drinking water, and aquatic environments. However, there is still a long way to go before MNMs can be used to track pollution in water-based environments. Recently, there has been a lot of interest in the use of machine learning to simulate potentially hazardous water quality issues. Research studies on the exploration of intricate statistical or mathematical relationships between magnetic attributes and other hazardous substances in water-related resources, in particular utilizing MNMs as NBSs, are still being evaluated, despite recent progress in the utilization of MNMs for managing potentially harmful chemicals, particularly HMs and dyes. Given this context, the use of intelligent devices, such as magnetic NBSs, for the monitoring of toxic water-based contaminants would be beneficial. By using predictive models, environmental scientists and engineers could potentially be spared considerable financial, time, and energy costs, as well as space requirements.

Even though MGS made of various materials are effective, there is still more research to be done on the gas-detecting capabilities of a number of newly developed advanced magnetic 2D materials before they can be used for more real-world applications. For instance, applications for MGS devices have still not been explored for black phosphorus, MXenes, or other classes of transition-metal dichalcogenides (TMDs). These 2D materials can have their magnetic characteristics tuned for use in gas detection devices by doping and altering them with magnetic composites. The characteristics of these substances can be additionally tuned to achieve improved MGS efficiency by varying the total number of layers and altering the 2D components using various techniques such as alloying, intercalation, defect and vacancy technology, adjusting the components in the x, y, and z directions, and creating heterostructures of 2D materials that possess various orientations. To motivate attempts to tackle issues including stability, specificity, velocity, multifunctionality, low power, and compatible operation in the presence of interference, more research is required. The last few years have seen the emergence of piezotronic and piezophototronic effects in GS research, which, with the right selection of MNMs, can be utilized in MGS systems. An additional essential field of research for MGS devices is the development of adaptable, practical, and self-sustaining GS systems. Utilizing environmentally friendly substances with high mechanical adaptability in this field of study will open up novel perspectives and possibilities for practical systems, especially in the NT/BNT domains. In this regard, it is necessary to investigate innovative and developing materials in MGS device configurations and technologies.

References

[1] Anani A, Adama K, Ukhurebor K, Habib A, Abanihi V and Pal K 2023 Application of nanofibrous protein for the purification of contaminated water as a next generational sorption technology: a review *Nanotechnology* **34** 1–18

[2] Ukhurebor K, Hossain I, Pal K, Jokthan G, Osang F, Ebrima F and Katal D 2023 Applications and contemporary issues with adsorption for water monitoring and remediation: a facile review *Top. Catal.* **67** 140–55

[3] Hojjati-Najafabadi A, Mansoorianfar M, Liang T, Shahin K and Karimi-Maleh H 2022 A review on magnetic sensors for monitoring of hazardous pollutants in water resources *Sci. Total Environ.* **824** 153844

[4] Wang M, Jin L, Hang-Mei W-N C L P, Zhao X, Chen H, Pan W, Liu H and Li S 2024 Advancements in magnetic nanoparticle-based biosensors for point-of-care testing *Front. Bioeng. Biotechnol.* **12** 1393789

[5] Zhang K, Song X, Liu M, Chen M, Li J and Han J 2023 Review on the use of magnetic nanoparticles in the detection of environmental pollutants *Water* **15** 3077

[6] Rathi B, Kumar P and Vo D-V 2021 Critical review on hazardous pollutants in water environment: occurrence, monitoring, fate, removal technologies and risk assessment *Sci. Total Environ.* **797** 149134

[7] Ukhurebor K, Ngonso B, Egielewa P, Cirella G, Akinsehinde B and Balogun V 2023 Petroleum spills and the communicative response from petroleum agencies and companies: impact assessment from the Niger Delta Region of Nigeria *Extract. Indus. Soc.* **15** 101331

[8] Eleryan A, Aigbe U, Ukhurebor K, Onyancha R, Hassaan M, Elkatory M, Ragab S, Osibote O, Kusuma H and El Nemr A 2023 Adsorption of Direct Blue 106 dye using zinc oxide nanoparticles prepared via green synthesis technique *Environ. Sci. Pollut. Res.* **30** 69666–82

[9] Aidonojie P, Okuonghae N and Ukhurebor K 2022 The legal rights and challenges of Covid-19 patients accessing private healthcare in Nigeria *Bestuur* **10** 183–97

[10] Neolaka Y *et al* 2022 Synthesis of zinc (II)-natural zeolite mordenite type as a drug carrier for ibuprofen: drug release kinetic modeling and cytotoxicity study *Res. Chem.* **4** 100578

[11] Ukhurebor K, Hussain A, Adetunji C, Aigbe U, Onyancha R and Abifarin O 2021 Environmental implications of petroleum spillages in the Niger Delta Region of Nigeria: a review *J. Environ. Manage.* **293** 112872

[12] Aigbe U, Ukhurebor K, Onyancha R, Okundaye B, Pal K, Osibote O, Esiekpe E, Kusuma H and Darmokoesoemo H 2022 A facile review on the sorption of heavy metals and dyes using bionanocomposites *Adsorp. Sci. Technol.* **2022** 8030175

[13] Neolaka Y, Riwu A, Aigbe U, Ukhurebor K, Onyancha R, Darmokoesoemo H and Kusuma H 2023 Potential of activated carbon from various sources as a low-cost adsorbent to remove heavy metals and synthetic dyes *Results in Chemistry* **5** 100711

[14] Hatami M, Sharifi A, Karimi-Maleh H, Agheli H and Karaman C 2021 Simultaneous improvements in antibacterial and flame retardant properties of PET by use of bionano-technology for fabrication of high performance PET bionanocomposites *Environ. Res.* **206** 112281

[15] Karaman C, Karaman O, Show P-L, Orooji Y and Karimi-Maleh H 2021 Utilization of a double-cross-linked amino-functionalized three-dimensional graphene networks as a mono-lithic adsorbent for methyl orange removal: equilibrium, kinetics, thermodynamics and artificial neutral network modeling *Environ. Res.* **207** 112156

[16] Karimi-Maleh H, Ayati A, Ghanbari S, Orooji Y, Tanhaei B, Karimi F *et al* 2021 Recent advances in removal techniques of Cr (VI) toxic ion from aqueous solution: a compre-hensive review *J. Mol. Liq.* **329** 115062

[17] Ambashta R and Sillanpää M 2010 Water purification using magnetic assistance: a review *J. Hazard. Mater.* **180** 38–49

[18] Azizi A, Shahhoseini F and Bottaro C 2020 Magnetic molecularly imprinted polymers prepared by reversible addition fragmentation chain transfer polymerization for dispersive solid phase extraction of polycyclic aromatic hydrocarbons in water *J. Chromatogr. A* **1610** 460534

[19] Berkani M, Vasseghian Y, Le V, Dragoi E-N and Mousavi K A 2021 and, 'The Fenton-like reaction for arsenic removal from groundwater: health risk assessment *Environ. Res.* **202** 111698

[20] Davar F, Shayan N, Hojjati-Najafabadi A, Sabaghi V and Hasani S 2017 Development of ZrO_2–MgO nanocomposite powders by the modified sol-gel method *Int. J. Appl. Ceram. Technol.* **14** 211–9

[21] Mahmoodi S, Hassan D, Hojjati-Najafabadi A, Li W, Liao L, Moshayedi A *et al* 2020 Quality enhancement of copper oxide thin film synthesized under elevated gravity accel-eration by two-axis spin coating *Ceram. Int.* **46** 7421–9

[22] Singh P, Sharma K, Hasija V, Sharma V, Sharma S, Raizada P *et al* 2019 Systematic review on applicability of magnetic iron oxides–integrated photocatalysts for degradation of organic pollutants in water *Mater. Today Chem.* **14** 100186

[23] Sharma M, Kalita P, Senapati K and Garg A 2018 *Emerging Pollutants— Some Strategies for the Quality Preservation of Our Environment* (London: IntechOpen)

[24] Sharma S, Yadav R and Singh A 2021 Presence of medicinal materials in drinking water: a review *Mater. Today Proc.* **2021** 1067–72

[25] Berkani M, Smaali A, Kadmi Y, Almomani F, Vasseghian Y, Lakhdari N *et al* 2022 Photocatalytic degradation of penicillin G in aqueous solutions: kinetic, degradation pathway, and microbioassays assessment *J. Hazard. Mater.* **421** 126719

[26] Berkani M, Vasseghian Y, Le V, Dragoi E-N and Mousavi K A 2021 The Fenton-like reaction for arsenic removal from groundwater: health risk assessment *Environ. Res.* **202** 111698

[27] Hojjati-Najafabadi A, Davar F, Enteshari Z and Hosseini-Koupaei M 2021 Antibacterial and photocatalytic behaviour of green synthesis of $Zn_{0.95}Ag_{0.05}O$ nanoparticles using herbal medicine extract *Ceram. Int.* **47** 31617–24

[28] Li X, Yang B, Yang J, Fan Y, Qian X and Li H 2021 Magnetic properties and its application in the prediction of potentially toxic elements in aquatic products by machine learning *Sci. Total Environ.* **783** 147083

[29] Torkian N, Bahrami A, Hosseini-Abari A, Momeni M, Abdolkarimi-Mahabadi M, Bayat A *et al* 2022 Synthesis and characterization of Ag-ion-exchanged zeolite/TiO_2 nano-composites for antibacterial applications and photocatalytic degradation of antibiotics *Environ. Res.* **207** 112157

[30] Abdi R 2020 Determining caffeic acid in food samples using a voltammetric sensor amplified by Fe_3O_4 nanoparticles and room temperature ionic liquid *Int. J. Electrochem. Sci.* **15** 2539–48

[31] Cui K, Yang T, Chen Y, Weerasooriya R, Li G, Zhou K *et al* 2021 Magnetic recyclable heterogeneous catalyst Fe_3O_4/g-C_3N_4 for tetracycline hydrochloride degradation via photo-Fenton process under visible light *Environ. Technol.* **43** 3341–54

[32] Liang Y, Li Z, Shi P, Ling C, Chen X, Zhou Q *et al* 2019 Performance of a novel magnetic solid-phase-extraction microsphere and its application in the detection of organic micro-pollutants in the Huai River China *Environ. Pollut.* **252** 196–204

[33] Maharana M and Sen S 2021 Magnetic zeolite: a green reusable adsorbent in wastewater treatment *Mater. Today Proc.* **47** 1490–5

[34] Nguyen T, Nguyen V-C, Phan T, Le V, Vasseghian Y, Trubitsyn M *et al* 2022 Novel biogenic silver and gold nanoparticles for multifunctional applications: green synthesis, catalytic and antibacterial activity, and colorimetric detection of Fe(III) ions *Chemosphere* **287** 132271

[35] Orooji Y, Tanhaei B, Ayati A, Tabrizi S, Alizadeh M, Bamoharram F *et al* 2021 Heterogeneous UV-Switchable Au nanoparticles decorated tungstophosphoric acid/TiO_2 for efficient photocatalytic degradation process *Chemosphere* **281** 130795

[36] Mustafa G, Shahzeb Khan M, Asif M, Ullah A, Khan I and Ullah I 2021 Water pollutants and nanosensors *Aquananotechnology* ed K A Abd-Elsalam and M Zahid (Amsterdam: Elsevier) pp 105–33

[37] Rhouati A, Berkani M, Vasseghian Y and Golzadeh N 2022 MXene-based electrochemical sensors for detection of environmental pollutants: a comprehensive review *Chemosphere* **291** 132921

[38] Xuan X, Yoon H and Park J 2018 A wearable electrochemical glucose sensor based on simple and low-cost fabrication supported micro-patterned reduced graphene oxide nano-composite electrode on flexible substrate *Biosens. Bioelectron.* **109** 75–82

[39] Manoochehri M and Naghibzadeh L 2016 A nanocomposite based on dipyridylamine functionalized magnetic multiwalled carbon nanotubes for separation and preconcentration of toxic elements in black tea leaves and drinking water *Food Anal. Methods* **10** 1777–86

[40] Karimian H, Li Q, Li C, Chen G, Mo Y, Wu C *et al* 2019 Spatio-temporal variation of wind influence on distribution of fine particulate matter and its precursor gases *Atmos. Pollut. Res.* **10** 53–64

[41] Akhoondi A, Feleni U, Bethi B, Idris A and Hojjati-Najafabadi A 2021 Advances in metal-based vanadate compound photocatalysts: synthesis, properties and applications *Synth. Sinter.* **1** 151–68

[42] Feng X, Yu Z, Sun Y, Long R, Shan M, Li X *et al* 2021 Review MXenes as a new type of nanomaterial for environmental applications in the photocatalytic degradation of water pollutants *Ceram. Int.* **47** 7321–43

[43] Jun B-M, Kim S, Heo J, Park C, Her N, Jang M *et al* 2019 Review of MXenes as new nanomaterials for energy storage/delivery and selected environmental applications *Nano Res.* **12** 471–87

[44] Mansoorianfar M, Khataee A, Riahi Z, Shahin K, Asadnia M, Razmjou A *et al* 2020 Scalable fabrication of tunable titanium nanotubes via sonoelectrochemical process for biomedical applications *Ultrason. Sonochem.* **64** 104783

[45] Vasseghian Y, Almomani F, Le V, Moradi M and Dragoi E-N 2022 Decontamination of toxic Malathion pesticide in aqueous solutions by Fenton-based processes: degradation pathway, toxicity assessment and health risk assessment *J. Hazard. Mater.* **423** 127016

[46] Khan F, Mubarak N, Tan Y, Khalid M, Karri R, Walvekar R *et al* 2021 A comprehensive review on magnetic carbon nanotubes and carbon nanotube-based buckypaper for removal of heavy metals and dyes *J. Hazard. Mater.* **413** 125375

[47] Vasseghian Y, Dragoi E-N, Almomani F and Le V 2022 Graphene-based materials for metronidazole degradation: a comprehensive review *Chemosphere* **286** 131727

[48] Onyancha R, Aigbe U, Ukhurebor K, Kusuma H, Darmokoesoemo H, Osibote O and Pal K 2022 Influence of magnetism-mediated potentialities of recyclable adsorbents of heavy metal ions from aqueous solutions—an organized review *Res. Chem.* **4** 100452

[49] Ahmadian-Fard-Fini S, Ghanbari D, Amiri O and Salavati-Niasari M 2021 Green sonochemistry assisted synthesis of hollow magnetic and photoluminescent $MgFe_2O_4$–carbon dot nanocomposite as a sensor for toxic Ni(ii), Cd(ii) and Hg(ii) ions and bacteria *RSC Adv.* **11** 22805

[50] Bagheri N, Habibi B, Khataee A and Hassanzadeh J 2019 Application of surface molecular imprinted magnetic graphene oxide and high performance mimetic behavior of bi-metal ZnCo MOF for determination of atropine in human serum *Talanta* **201** 286–94

[51] Cheraghi A, Davar F, Homayoonfal M and Hojjati-Najafabadi A 2021 Effect of lemon juice on microstructure, phase changes, and magnetic performance of $CoFe_2O_4$ nanoparticles and their use on release of anti-cancer drugs *Ceram. Int.* **47** 20210–9

[52] Mateus A, Torres J, Marimon-Bolivar W and Pulgarin L 2021 Implementation of magnetic bentonite in food industry wastewater treatment for reuse in agricultural irrigation *Water Resour. Ind.* **26** 100154

[53] You J, Wang L, Zhao Y and Bao W 2021 A review of amino-functionalized magnetic nanoparticles for water treatment: features and prospects *J. Clean. Prod.* **281** 124668

[54] Vinosha P, Manikandan A, Preetha A, Dinesh A, Slimani Y, Almessiere M *et al* 2021 Review on recent advances of synthesis, magnetic properties, and water treatment applications of cobalt ferrite nanoparticles and nanocomposites *J. Supercond. Nov. Magn.* **34** 995–1018

[55] Rajendran S, Priya T, Khoo K, Hoang T, Ng H-S, Munawaroh H *et al* 2022 A critical review on various remediation approaches for heavy metal contaminants removal from contaminated soils *Chemosphere* **287** 132369

[56] Tamjidi S, Esmaeili H and Kamyab M B 2019 Application of magnetic adsorbents for removal of heavy metals from wastewater: a review study *Mater. Res. Express* **6** 102004

[57] Nithya R, Thirunavukkarasu A, Sathya A and Sivashankar R 2021 Magnetic materials and magnetic separation of dyes from aqueous solutions: a review *Environ. Chem. Lett.* **19** 1275–94

[58] Priya A, Gnanasekaran L, Rajendran S, Qin J and Vasseghian Y 2021 Occurrences and removal of pharmaceutical and personal care products from aquatic systems using advanced treatment—a review *Environ. Res.* **204** 112298

[59] Shinde P and Rout C 2021 Magnetic gas sensing: working principles and recent developments *Nanoscale Adv.* **3** 1551–68

[60] Almomani F, Bhosale R, Khraisheh M and Almomani T 2020 Heavy metal ions removal from industrial wastewater using magnetic nanoparticles (MNP) *Appl. Surf. Sci.* **506** 144924

[61] Santhosh C, Daneshvar E, Kollu P, Peräniemi S, Grace A and Bhatnagar A 2017 Magnetic SiO_2 @$CoFe_2O_4$ nanoparticles decorated on graphene oxide as efficient adsorbents for the removal of anionic pollutants from water *Chem. Eng. J.* **322** 472–87

[62] Zhang H, Li M, Zhu C, Tang Q, Kang P and Cao J 2020 Preparation of magnetic α-Fe_2O_3/ $ZnFe_2O_4$@ Ti_3C_2 MXene with excellent photocatalytic performance *Ceram. Int.* **46** 81–8

[63] Turner A, Karube I and Wilson G 1987 *Biosensors Fundamentals and Applications* (Oxford: Oxford University Press)

[64] Ukhurebor K, Odesanya I, Tyokighir S, Kerry R, Olayinka A and Bobadoye A 2020 Wireless sensor networks: applications and challenges *Wireless Sensor Networks—Design, Deployment and Applications* (London: IntechOpen)

[65] Onyancha R, Ukhurebor K, Aigbe U, Osibote O, Kusuma H, Darmokoesoemo H and Balogun V 2021 A systematic review on the detection and monitoring of toxic gases using carbon nanotube-based biosensors *Sens. Bio-Sens. Res.* **34** 100463

[66] Singh K, Nayak V, Adetunji C, Ukhurebor K, Singh J and Singh R 2022 Introduction to nanobiosensors *Nanobiosensors for Environmental Monitoring: Fundamentals and Application* (Berlin: Springer Nature) pp 1–17

[67] Ukhurebor K, Adetunji C, Bobadoye A, Aigbe U, Onyancha R, Siloko I, Emegha J, Okocha G and Abiodun I 2021 Bionanomaterials for biosensor technology *Bionanomaterials: Fundamentals and Biomedical Applications* (Bristol: Institute of Physics Publishing) pp 1–22

[68] Ukhurebor K 2020 The role of biosensor in climate smart organic agriculture towards agricultural and environmental sustainability *Agrometeorology* (London: IntechOpen) p 8

[69] Ndunagu J, Ukhurebor K, Akaaza M and Onyancha R 2022 Development of a wireless sensor network and iot-based smart irrigation system *Appl. Environ. Soil Sci.* **2022** 7678570

[70] Ukhurebor K *et al* 2021 Developments, utilization and applications of nanobiosensors for environmental sustainability and safety *Bionanomaterials for Environmental and Agricultural Applications* (Bristol: Institute of Physics Publishing) p 14

[71] Kaur H, Raveendran A, Athwal S, Verma M, Mutreja V, Ukhurebor K and Kusuma H 2023 Assessment of diabetes biomarker monitoring via novel biosensor activity *Res. Chem.* **5** 1–9

[72] Paladhi A, Manohar M, Pal K, Vallinayagam S, Packirisamy A, Bashreer V, S.R N and Ukhurebor K 2022 Novel electrochemical biosensor key significance of smart intelligence (IoMT & IoHT) of COVID-19 virus control management *Process Biochem.* **122** 105–9

[73] Ukhurebor K, Onyancha R, Aigbe U, UK-Eghonghon G, Kerry R, Kusuma H, Darmokoesoemo H, Osibote O and Balogun V 2022 A methodical review on the applications and potentialities of using nanobiosensors for diseases diagnosis *Bio Med Res. Int.* **2022** 1682502

[74] Kerry R *et al* 2021 A comprehensive review on the applications of nano-biosensor based approaches for non-communicable and communicable disease detection *Biomater. Sci.* **9** 3576–602

[75] Singh R, Ukhurebor K, Singh J, Adetunji C and Singh K 2022 *Nanobiosensors for Environmental Monitoring—Fundamentals and Application* (Berlin: Springer Nature)

[76] Aidonojie P, Ukhurebor K, Masajuwa F, Imoisi S, Edetalehn O and Nwazi J 2022 Legal implications of nanobiosensors concerning environmental monitoring *Nanobiosensors for Environmental Monitoring: Fundamentals and Application* (Berlin: Springer Nature) pp 439–58

[77] Ukhurebor K and Adetunji C 2021 Relevance of biosensor in climate smart organic agriculture and their role in environmental sustainability: what has been done and what we need to do *Biosensors in Agriculture: Recent Trends and Future Perspective* (Berlin: Springer Nature) pp 115–36

[78] Otun K, Azeez I, Ama O, Anku W, Aigbe U, Ukhurebor K and Onyancha R 2022 Sensing the presence of inorganic ions in water: the use of electrochemical sensors *Modified Nanomaterials for Environmental Applications—Electrochemical Synthesis, Characterization and Properties* (Berlin: Springer Nature) pp 65–89

[79] Ciprian R *et al* 2016 New strategy for magnetic gas sensing *RSC Adv.* **6** 83399–405

[80] Lueng C, Lupo P, Schefer T, Metaxas P, Adeyeye A and Kostylev M 2019 Sensitivity of ferromagnetic resonance in PdCo alloyed films to hydrogen gas *Int. J. Hydrogen Energy* **44** 7715–24

[81] Ignatyeva D, Knyazev G, Kapralov G, Dietler G, Sekatskii S and Belotelov V 2016 Magneto-optical plasmonic heterostructure with ultranarrow resonance for sensing applications *Sci. Rep.* **6** 28077

[82] Hsu C-C, Chang P-C, Chen Y-H, Liu C-M, Wu C-T, Yen H-W and Lin W-C 2018 Reversible 90-degree rotation of Fe magnetic moment using hydrogen *Sci. Rep.* **8** 1–2

[83] Matatagui D, Kolokoltsev O, Qureshi N, Mejía-Uriarte E and J.M S 2015 A magnonic gas sensor based on magnetic nanoparticles *Nanoscale* **7** 9607–13

[84] Moseley P 1992 Materials selection for semiconductor gas sensors *Sensors Actuators* B **6** 149–56

[85] Taylor R and Schultz J 1996 *Handbook of Chemical and Biological Sensors* (New York: CRC Press)

[86] Punnoose A, Reddy K, Hays J, Thurber A and Engelhard M 2006 Magnetic gas sensing using a dilute magnetic semiconductor *Appl. Phys. Lett.* **89** 112509

[87] Hanley C, Thurber A, Hanna C, Punnoose A, Zhang J and Wingett D 2009 The influences of cell type and ZnO nanoparticle size on immune cell cytotoxicity and cytokine induction *Nanoscale Res. Lett.* **4** 1409–20

[88] Mustafa G, Shahzeb Khan M, Asif M, Ullah A, Khan I and Ullah I 2021 Water pollutants and nanosensors *Aquananotechnology* (Amsterdam: Elsevier) pp 105–33

[89] Pooja D, Kumar P, Singh P and Patil S 2020 *Sensors in Water Pollutants Monitoring: Role of Material* (Berlin: Springer Nature)

[90] Rasheed T, Bilal M, Nabeel F, Adeel M and Iqbal H 2019 Environmentally-related contaminants of high concern: lential sources and analytical modalities for detection, quantification, and treatment *Environ. Int.* **122** 52–66

[91] Aylaz G, Kuhn J, Lau E, Yeung C-C, Roy V, Duman M and Yiu H 2021 Recent developments on magnetic molecular imprinted polymers (MMIPs) for sensing, capturing, and monitoring pharmaceutical and agricultural pollutants *J. Chem. Technol. Biotechnol.* **96** 1151–60

[92] Karimi-Maleh H, Fakude C, Mabuba N, Peleyeju G and Arotiba O 2019 The determination of 2-phenylphenol in the presence of 4-chlorophenol using nano-Fe_3O_4/ionic liquid paste electrode as an electrochemical sensor *J. Colloid Interface Sci.* **554** 603–10

[93] Karimi-Maleh H, Ayati A, Davoodi R, Tanhaei B, Karimi F, Malekmohammadi S *et al* 2021 Recent advances in using of chitosan-based adsorbents for removal of pharmaceutical contaminants: a review *J. Clean. Prod.* **291** 125880

[94] Vijayalakshmi M and Deepa B 2011 Impact of industrial effluent treatment on profitability of small and medium enterprises: a case study of Hyderabad bulk drug industry *Entrepreneurship and SMEs: Building Competencies* (New Delhi: Macmillan)

[95] Chen J, Alcaine S, Jiang Z, Rotello V and Nugen S 2015 Detection of *Escherichia coli* in drinking water using T7 bacteriophage-conjugated magnetic probe *Anal. Chem.* **87** 8977–84

[96] Castillo-Torres K *et al* 2019 Rapid isolation of *Escherichia coli* from water samples using magnetic microdiscs *Sensors Actuators* B **291** 58–66

[97] Jia F, Xu L, Yan W, Wu W, Yu Q, Tian X *et al* 2017 A magnetic relaxation switch aptasensor for the rapid detection of *Pseudomonas aeruginosa* using superparamagnetic nanoparticles *Microchim. Acta* **184** 1539–45

[98] Alsaiari N, Katubi K, Alzahrani F, Siddeeg S and Tahoon M 2021 The application of nanomaterials for the electrochemical detection of antibiotics: a review *Micromachines* **12** 308

[99] Canciu A, Tertis M, Hosu O, Cernat A, Cristea C and Graur F 2021 Modern analytical techniques for detection of bacteria in surface and wastewaters *Sustainability* **13** 7229

[100] Suaifan G, Alhogail S and Zourob M 2017 Paper-based magnetic nanoparticle-peptide probe for rapid and quantitative colorimetric detection of *Escherichia coli* O157:H7 *Biosens. Bioelectron.* **92** 702–8

[101] Yao L, Wang L, Huang F, Cai G, Xi X and Lin J 2018 A microfluidic impedance biosensor based on immunomagnetic separation and urease catalysis for continuous-flow detection of *E. coli* O157:H7 *Sensors Actuators* B **259** 1013–21

[102] Zhao M, Wang J and Lian Z 2021 Fluorescence assay of oxytetracycline in seawater after selective capture using magnetic molecularly imprinted nanoparticles *Mar. Pollut. Bull.* **163** 111962

[103] Qin S, Su L, Wang P and Gao Y 2015 Rapid and selective extraction of multiple sulphonamides from aqueous samples based on Fe_3O_4–chitosan molecularly imprinted polymers *Anal. Methods* **7** 8704–13

[104] Zamora-Gálvez A, Ait-Lahcen A, Mercante L, Morales-Narváez E, Amine A and Merkoçi A 2016 Molecularly imprinted polymer-decorated magnetite nanoparticles for selective sulfonamide detection *Anal. Chem.* **88** 3578–84

[105] Umapathi R, Ghoreishian S, Sonwal S, Rani G and Huh Y 2022 Portable electrochemical sensing methodologies for on-site detection of pesticide residues in fruits and vegetables *Coord. Chem. Rev.* **453** 214305

[106] Umapathi R, Park B, Sonwal S, Rani G, Cho Y and Huh Y 2022 Advances in optical sensing strategies for the on-site detection of pesticides in agricultural foods *Trends Food Sci. Technol.* **119** 69–89

[107] Umapathi R, Sonwal S, Lee M, Mohana Rani G, Lee E-S, Jeon T-J *et al* 2021 Colorimetric based on-site sensing strategies for the rapid detection of pesticides in agricultural foods: New horizons, perspectives, and challenges *Coord. Chem. Rev.* **446** 214061

[108] Yang G, He Z, Liu X, Liu C, Zhan J, Liu D *et al* 2016 Polymer-coated magnetic nanospheres for preconcentration of organochlorine and pyrethroid pesticides prior to their determination by gas chromatography with electron capture detection *Microchim. Acta* **183** 1187–94

[109] Akbarzade S, Chamsaz M, Rounaghi G and Ghorbani M 2018 Zero valent fe-reduced graphene oxide quantum dots as a novel magnetic dispersive solid phase microextraction sorbent for extraction of organophosphorus pesticides in real water and fruit juice samples prior to analysis by gas chromatography-mass spectrum *Anal. Bioanal. Chem.* **410** 429–39

[110] Rashidi Nodeh H, Wan Ibrahim W, Kamboh M and Sanagi M 2017 New magnetic graphene-based inorganic–organic sol-gel hybrid nanocomposite for simultaneous analysis of polar and non-polar organophosphorus pesticides from water samples using solidphase extraction *Chemosphere* **166** 21–30

[111] Xie J, Liu T, Song G, Hu Y and Deng C 2013 Simultaneous analysis of organophosphorus pesticides in water by magnetic solid-phase extraction coupled with GC–MS *Chromatographia* **76** 535–40

[112] Nasiri M, Ahmadzadeh H and Amiri A 2021 Organophosphorus pesticides extraction with polyvinyl alcohol coated magnetic graphene oxide particles and analysis by gas chroma-tography–mass spectrometry: application to apple juice and environmental water *Talanta* **227** 122078

[113] Ben Attig J, Latrous L, Zougagh M and Ríos Á 2021 Ionic liquid and magnetic multiwalled carbon nanotubes for extraction of N-methylcarbamate pesticides from water samples prior their determination by capillary electrophoresis *Talanta* **226** 122106

[114] Lu H and Liu W 2016 Distribution characteristics of organochlorine pesticides in soil, water, and sediment from the Bahe River China *Environ. Forensics* **17** 80–6

[115] Luo D, Pu Y, Tian H, Cheng J, Zhou T, Tao Y *et al* 2016 Concentrations of organo-chlorine pesticides in umbilical cord blood and related lifestyle and dietary intake factors among pregnant women of the Huaihe River Basin in China *Environ. Int.* **92–93** 276–83

[116] Wang H, Qu B, Liu H, Ding J and Ren N 2018 Analysis of organochlorine pesticides in surface water of the Songhua River using magnetoliposomes as adsorbents coupled with GCMS/MS detection *Sci. Total Environ.* **618** 70–9

[117] Zhou Q, Wu Y, Sun Y, Sheng X, Tong Y, Guo J *et al* 2021 Magnetic polyamidoamine dendrimers for magnetic separation and sensitive determination of organochlorine pesti-cides from water samples by high-performance liquid chromatography *J. Environ. Sci.* **102** 64–73

[118] Borah P, Kumar M and Devi P 2020 Types of inorganic pollutants: metals/metalloids, acids, and organic forms *Inorganic Pollutants in Water* (Amsterdam: Elsevier)

[119] Soleimani M, Ghasemi J, Mohammadi Ziarani G, Karimi-Maleh H and Badiei A 2021 Photocatalytic degradation of organic pollutants, viral and bacterial pathogens using titania nanoparticles *Inorg. Chem. Commun.* **130** 108688

[120] Zhang L, Guo Y, Hao R, Shi Y, You H, Nan H *et al* 2021 Ultra-rapid and highly efficient enrichment of organic pollutants via magnetic mesoporous nanosponge for ultrasensitive nanosensors *Nat. Commun.* **12** 6849

[121] Wang M, Fu Q, Zhang K, Wan Y, Wang L, Gao M *et al* 2019 A magnetic and carbon dot based molecularly imprinted composite for fluorometric detection of 2,4,6-trinitrophenol *Mikrochim Acta* **186** 86

[122] Xu M, An Y, Wang Q, Wang J, Hao L, Wang C *et al* 2021 Construction of hydroxyl functionalized magnetic porous organic framework for the effective detection of organic micropollutants in water, drink and cucumber samples *J. Hazard. Mater.* **412** 125307

[123] Zhao M, Hou Z, Lian Z, Qin D and Ge C 2020 Direct extraction and detection of malachite green from marine sediments by magnetic nano-sized imprinted polymer coupled with spectrophotometric analysis *Mar. Pollut. Bull.* **158** 111363

[124] He J, Song G, Wang X, Zhou L and Li J 2022 Multifunctional magnetic Fe_3O_4/GO/Ag composite microspheres for SERS detection and catalytic degradation of methylene blue and ciprofloxacin *J. Alloys Compd.* **893** 162226

[125] Beitollahi H, Nejad F and Tajik S 2021 Application of magnetic nanomaterials as electrochemical sensors *Magnetic Nanomaterials in Analytical Chemistry* (Amsterdam: Elsevier) pp 269–301

[126] Sawan S, Maalouf R, Errachid A and Jaffrezic-Renault N 2020 Metal and metal oxide nanoparticles in the voltammetric detection of heavy metals: a review *TrAC, Trends Anal. Chem.* **131** 116014

[127] Ahmadian-Fard-Fini S, Ghanbari D, Amiri O and Salavati-Niasari M 2021 Green sonochemistry assisted synthesis of hollow magnetic and photoluminescent $MgFe_2O_4$– carbon dot nanocomposite as a sensor for toxic Ni(II), Cd(II) and Hg(II) ions and bacteria. RS *RSC Adv.* **11** 22805–11

[128] Kong Y, Wu T, Wu D, Zhang Y, Wang Y, Du B *et al* 2018 An electrochemical sensor based on Fe_3O_4@PANI nanocomposites for sensitive detection of Pb^{2+} and Cd^{2+} *Anal. Methods* **10** 4784–92

[129] Qiu B, Xie L, Zeng J, Liu T, Yan M, Zhou S *et al* 2021 Interfacially super-assembled asymmetric and h2o2 sensitive multilayer-sandwich magnetic mesoporous silica nano-motors for detecting and removing heavy metal ions *Adv. Funct. Mater.* **31** 2010694

[130] Chimezie A, Hajian R, Yusof N, Woi P and Shams N 2017 Fabrication of reduced graphene oxide-magnetic nanocomposite (rGO-Fe_3O_4) as an electrochemical sensor for trace determination of As(III) in water resources *J. Electroanal. Chem.* **796** 33–42

[131] Baghayeri M, Alinezhad H, Fayazi M, Tarahomi M, Ghanei-Motlagh R and Maleki B 2019 A novel electrochemical sensor based on a glassy carbon electrode modified with dendrimer functionalized magnetic graphene oxide for simultaneous determination of trace Pb(II) and Cd(II) *Electrochim. Acta* **312** 80–8

[132] Mohammadi S, Taher M, Beitollahi H and Naghizadeh M 2019 Sensitive voltammetric determination of cadmium at a carbon nanotubes/Fe_3O_4/eggshell composites modified carbon paste electrode *Environ. Nanotechnol. Monit. Manag.* **12** 100241

[133] Wang L, Jiang X, Su S, Rao J, Ren Z, Lei T *et al* 2021 A thiol and magnetic polymerbased electrochemical sensor for on-site simultaneous detection of lead and copper in water *Microchem. J.* **168** 106493

[134] Fan C, Chen L, Jiang R, Ye J, Li H, Shi Y *et al* 2021 $ZnFe_2O_4$ nanoparticles for electrochemical determination of trace Hg(II), Pb(II), Cu(II), and glucose *ACS Appl. Nano Mater.* **4** 4026–36

[135] Yang Q, Dong Y, Qiu Y, Yang X, Cao H and Wu Y 2020 Design of functional magnetic nanocomposites for bioseparation *Colloids Surf. B: Biointerfaces* **191** 111014

[136] van de Walle A B *et al* 2020 Magnetic nanoparticles in regenerative medicine: what of their fate and impact in stem cells? *Mater. Today Nano* **11** 100084

[137] Bushra R, Ahmad M, Alam K, Farzad Seidi F, Qurtulen, Shakeel S, Song J, Jin Y and Huining Xiao H 2024 Recent advances in magnetic nanoparticles: key applications, environmental insights, and future strategies *Sustain. Mater. Technol.* **40** e00985

Uyiosa Osagie Aigbe and Kingsley Eghonghon Ukhurebor

Chapter 7

Challenges and perspectives for magnetic nanomaterials

The intrinsic characteristics of magnetic nanoparticles (MNPs) offer significant advantages for numerous technologies aimed at enhancing the standard of our environment. This book has focused on the bulk of the research into the use of MNMs and magnetic fields (MFs) for environmental applications. The main purpose of this book was to demonstrate how this potent research combination can significantly improve a variety of environmental applications, such as environmental sensing, the treatment of wastewater, and groundwater remediation. Consequently, this last chapter concludes the book with perspectives on and challenges of using MNMs and MFs for environmental applications.

7.1 Introduction

In view of their distinctive magnetic and structural characteristics, magnetic nanomaterials (MNMs) or MNPs are now becoming increasingly recognized for their wide range of uses in a variety of sectors in the environmental and medical/biomedical (BM) domains, as well as information storage, making them potential materials for various industrial sectors [1–6]. However, there are a number of challenges to their broad adoption, including the effects they have on the environment throughout their life cycle, health hazards associated with their toxicological profiles, and logistical hurdles in the recovery and recycling procedures [1, 3, 4, 6]. Their practical/applied integration into commercial and industrial sectors and applications is further impeded by challenges related to economic feasibility and operational restrictions [1–6].

The main scientific and technological issues that need to be resolved to improve MNM functionality and safety are outlined in this final chapter. These issues include

doi:10.1088/978-0-7503-6377-8ch7
7-1

improving environmental management strategies, creating safer coatings, and streamlining synthesis procedures for consistent particle control. This chapter's purview is expanded to include a review of MNMs' potential applications in the fields of BM procedures, the environment, catalysis, and data storage. MNMs play a significant role in environmental remediation, especially in the treatment of wastewater and pollutant sequestration. They are also essential in medicine administration, hyperthermia psychotherapy, and diagnostic imaging in the healthcare sector. Their increasing significance is further demonstrated by the fact that their catalytic qualities improve the efficiency and selectivity of industrial procedures. A life cycle assessment (LCA) that evaluates the environmental impact of MNMs from production to waste and identifies areas in need of further improvement is included in this evaluation. In order to provide insights into mitigation measures and evaluate possible dangers to human health, animal husbandry (livestock), and environmental systems, this chapter meticulously examines the toxicological assessments of MNMs. It covers recovery and recyclability issues, operational limitations, market challenges, and economic feasibility, offering comprehensive perspectives on the challenges associated with their commercial applications. To sum up, this review not only gathers the most recent information about the applications of MNMs but also initiates a study on their new functions, potential health risks, and efficient risk mitigation. The inclusion of economic and safety aspects improves our comprehension of MNMs' sustainable use. In turn, this provides a strategic roadmap for their further advancement and use that considers their effects on the environment and human health. Figure 7.1(a) shows the graphic representation of the bioseparation process of MNPs or MNMs [7], and figure 7.1(b) shows the applicative perspective on MNPs or MNMs in the BM domain [8], as adapted and reproduced from Bushra *et al* [1].

Consequently, this chapter provides an extensive review of the various applications of MNMs and assesses the environmental sustainability of their entire life cycle as part of the ongoing research into them. With a focus on producing safer protective coatings, enhancing synthesis, and enhancing environmental sustainability management approaches, this chapter seeks to critically examine three primary areas of interest. First, it emphasizes the potentially transformative benefits of the emerging capabilities of MNMs, such as data storage, the treatment and management of wastewater, catalysis, and food processing across many different sectors, placing particular attention on their roles within environmental sustainability management. Second, it conducts an extensive LCA of MNMs, evaluating their impact on the environment from production to disposal while highlighting areas for potential improvement. This chapter also delves into additional detail about possible hazards and methods for risk management related to MNMs, as well as the market opportunities, operational challenges, and potential future challenges these NMs might encounter. The advancement of research in this field has led to the recognition of MNMs as possible catalysts for paradigm shifts in other areas, hence broadening the scope of materials that combine innovation, science, and technology in relation to nanotechnology (NT) and bionanotechnology (BNT).

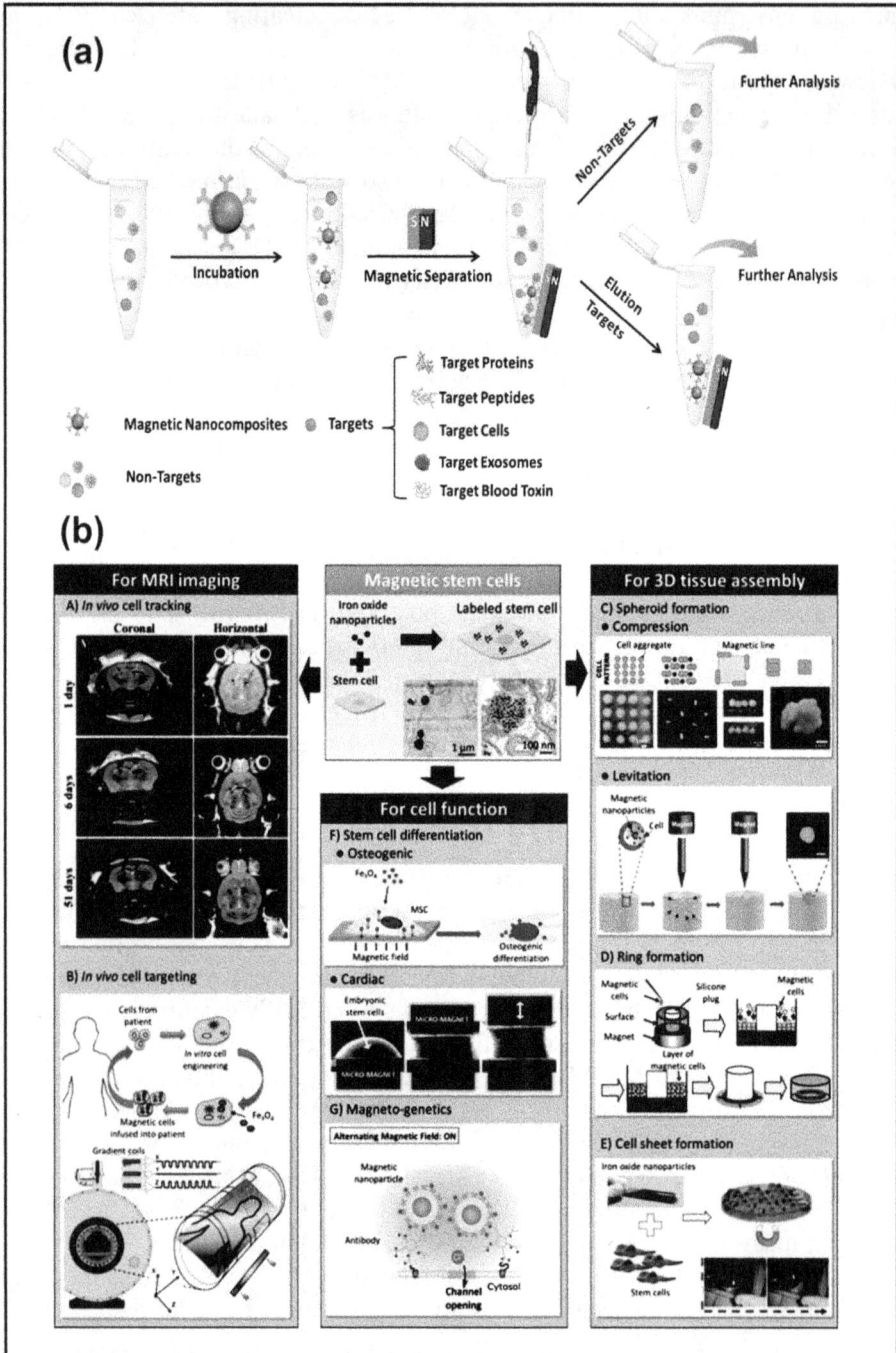

Figure 7.1. (a) Graphic representation of the bioseparation process of MNPs or MNMs. Reprinted from [7], Copyright (2020), with permission from Elsevier. (b) Applicative perspective on MNPs or MNMs in the BM domain. Reprinted from [8], Copyright (2024), with permission from Elsevier as adapted from [1], Copyright (2024), with permission from Elsevier.

7.2 Emerging capabilities and applications of magnetic nanomaterials

Recent advancements in technology and a sharp rise in scientific papers on the subject have made MNMs extremely significant in the fields of NT and BNT [1, 3–6]. They are very attractive because of their exceptional superparamagnetic qualities, unique shape and size features, outstanding surface area-to-volume ratio, as well as innate biocompatibility [1, 3–6]. Researchers from a variety of fields have expressed interest in these characteristics [1, 3–6]. This chapter delves into the diverse industrial uses of MNMs and provides an in-depth analysis of their possible roles in the storage of data, wastewater (effluent) management, and catalysis, as well as in the food industry, among other areas. The extensive investigation of various uses highlights the versatile nature and revolutionary potential of MNMs, elevating the field of materials science and technology in the process.

7.2.1 Applications of magnetic nanomaterials in data storage

Given that magnetic materials are capable of retaining data even in the absence of a power source, they are frequently utilized as nonvolatile memory storage materials. Floppy disks, magnetically recorded tapes, and magnetic stripes on credit cards constitute typical examples. Through the development of distinct magnetizable patterns inside magnetic materials, these devices store information [1]. Magnetic domains are microscopic magnetic areas on the surfaces of these types of substances with dimensions generally only a couple of micrometers across. They play a major role in facilitating the storage function. In general, these zones exhibit homogeneous magnetization levels. Since these magnetic compounds are polycrystalline, the function of each of the magnetic crystal grains becomes critical in this context. There are considered to be two types of magnetic media for storage: sequential access memory and random-access memory [1]. However, there may be situations in which it is difficult to differentiate between these two types. The access time, which is the typical time necessary to retrieve stored data, is a crucial quantity in this regard. The read head and the write head, both of which are electromagnetically highly sensitive information devices mounted atop moving devices, are the main parts of this system. These elements make up the vital components of the storage mechanism, along with the storage medium. At the write head, data in the form of sound energy is transformed into electrical and magnetic impulses to start the operational process. Using an MF and an electric field, this data is transmitted to the read head, allowing access for users. This kind of memory structure is referred to as magnetic recording in audio and video recording, but it is termed magnetic storage in information technology [9, 10].

However, the recent discovery of the spontaneously self-assembled state of magnetic FePt nanomaterials (NMs) or nanoparticles (NPs) on a substrate has led to the possibility of using MNMs in the storage of data, which is one of their most promising applications [11]. These NMs have been used to create magnetic storage materials that can achieve data storage densities of many terabits per square centimeter (terabits cm^{-2}). This development is predicated on the idea that, in order to boost the density of data storage, consistently oriented single ferromagnetic

NMs can be used in place of the bigger, irregularly oriented magnetic grains used in conventional media. There is a direct relationship between MNM storage density and particle size: more data may be stored by smaller MNMs. It is imperative, therefore, that these MNMs' magnetic moments remain stable over time. FePt alloyed NMs are specifically preferred for this application because of their magnetic 'hardness,' which enables them to robustly maintain their magnetic tendency even in the absence of an external MF [12]. However, when MNMs get smaller, thermal disorder may become greater than the inherent magnetic anisotropy of the components, which could result in unsteady magnetic moments. The 'superparamagnetic limit' is a phenomenon that places fundamental restrictions on increasing data density in storage applications. The use of exchange bias influence has only recently emerged as a viable way to bypass this constraint [13]. By introducing extra magnetic anisotropy, this allows MNMs to be reduced below the superparamagnetic threshold while preserving stable magnetization. Interestingly, hollow MNMs show an enhanced exchange bias influence, making them very attractive prospects for data storage applications [14]. Electrospun nanofiber mats have also been developed recently, and they have the potential to serve as novel building blocks for neuromorphic systems for information processing. These mats need to be magnetic, possibly electrically conductive, and responsive to outside stimuli in order to store and transfer data efficiently. The creation of these magnetic nanofiber mats, which are made of polymer nanofibers imbued with magnetic components enabling the transmission of data, has been described by Döpke *et al* [15]. For data storage purposes, these mats also include polymer beads (polyacrylonitrile) with increased concentrations of MNMs (magnetite, Fe_3O_4; Fe_2O_3/NiO). Magnetic nanofiber mats with different morphologies and different amounts of magnetically embedded beads were electrospun using a needle-free method, incorporating Fe_3O_4 and iron nickel oxide NMs as dopants.

MNM-based data storage requires dealing with a number of challenges, such as maintaining stability, streamlining read/write operations, and striking a balance between data density and dependability. Other challenges include ensuring that the system is resilient to the climate, scaling at a reasonable cost, increasing the speed of data transmission, strengthening durability, and preserving compatibility. MNM data storage needs ongoing study and advancement to prove its sustainability, competitiveness, and viability in the data storage market, despite its potential for high-capacity, long-lasting storage solutions. Furthermore, combining new machine learning techniques with nanophotonic-based optical data storage could result in novel approaches to writing and reading optical data that are precise, fast, reliable, and high resolution. This would make it easier to discover, develop, and improve NMs and nanostructures, giving them the newfound potential to expand the field of contemporary optical data storage based on nanophotonics [1, 16–22].

7.2.2 Applications of magnetic nanomaterials in the treatment and management of wastewater

Empirical evidence has demonstrated the effectiveness of MNMs in the treatment and management of wastewater, with several benefits due to the unique features they possess. These consist of inherent qualities including catalytic activity, high reactivity, greater mobility, and improved adsorption capabilities, in addition to magnetic, electrical, and optical properties. Together, these qualities make them a superior option compared to traditional treatment techniques [23].

MNMs have proven quite useful in handling different types of contaminants in water and wastewater, improving filtering efficiency, encouraging recycling, and reducing operating costs. In addition to lowering costs, using MNMs in nano-remediation techniques allows for *in situ* remediation, shorter treatment times, and remediation effectiveness that is close to 100.00%. Furthermore, combining MNMs with conventional wastewater treatment methods can result in better waste management and lower energy usage in thermal processes. Numerous MNMs with different sizes, chemical compositions, and structural arrangements have been created and used to clean up the environment. Adsorption, filtration, transformation, and catalysis are the four main processes that these NMs use to eliminate or break down pollutants [24–26]. For example, manganese iron oxycarbide with carboxylic functionalization (600 nm) efficiently produces reactive oxygen species (ROS) to break down the antimicrobial preservative butyl paraben [27]. To extract indole-3-butyric acid from aqueous solutions, magnetic layered double hydroxides, including Mg(II) and Al(III) composites, have been produced as alternative highly efficient adsorbents [28]. A magnetic chitosan-glutaraldehyde/ZnO/Fe_3O_4 NM is one of the further advancements. It has demonstrated a strong adsorption capability for removing Brilliant Blue R dye [29], reaching a removal rate of up to 176.60 mg g^{-1} in thirty-five minutes at 60.00 °C. Furthermore, quicker H_2O_2 breakdown has been achieved using copper (Cu)-doped Fe_3O_4 MNMs, which have been shown to remove rhodamine B from textile effluent with a proficiency of over 97.00% [30]. Pharmaceutical medicines were successfully removed by a MNM that was generated using a mix of solvothermal and solid-state dispersion processes [31]. Notably, even after five regeneration cycles, the NM retained good removal efficiency. In addition to effectively eliminating organic pollutants, MNMs are also effective in removing heavy metals (HMs) that are present in the environment as inorganic harmful substances, such as Cu, mercury (Hg), chromium (Cr), etc. For example, Pb^{2+} ions were removed from wastewater using ultrasound assistance and γ-Fe_2O_3 as an adsorbent [32], with a maximum adsorption capacity of 163.57 mg g^{-1} in just 240 s and adsorption kinetics that fitted the Langmuir isotherm model. Additionally, the removal of cyanate and HMs from samples of wastewater has been demonstrated by Rajput *et al* [33] and Ranjan *et al* [29]. As shown in figure 7.2(a), as reported by Sahoo *et al* [34] (adapted and reproduced from Bushra *et al* [1]), π–π interactions, electrostatic attraction, and hydrogen bonding are involved in the adsorption processes of thiacarbocyanine (TC) and methylene blue (MB) on a composite material consisting of graphene oxide (GO), graphitic carbon nitride (g-C_3N_4), and Fe_3O_4. The observed absorption behavior indicates a complex intermolecular interaction that includes hydrogen bridge creation, aromatic stacking, and charge-induced attraction. All of these processes contribute to the total adsorption process of the composite material.

Even though MNMs have an established track record of being effective in treating synthetic wastewater, further research in the field is urgently required to

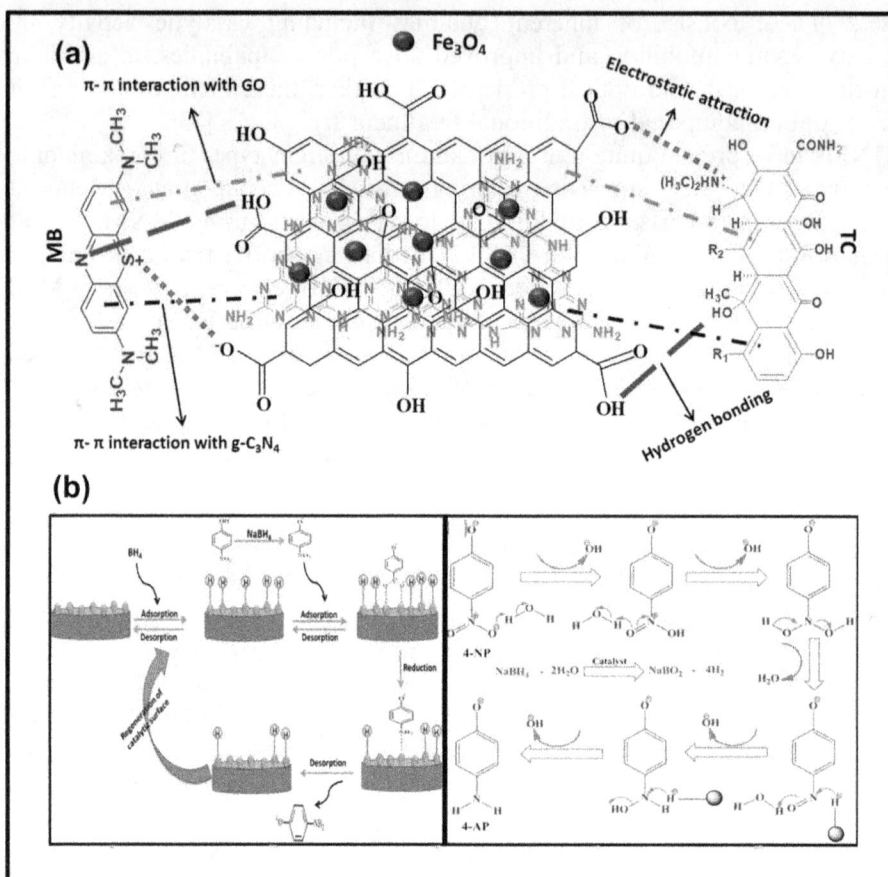

Figure 7.2. (a) Potential interactions between TC and MB and the surface of the GO/g-C$_3$N$_4$–Fe$_3$O$_4$ NM. (b) iron oxide nanoparticles decorated montmorillonite clay (MMTPFe) NM-catalyzed reduction of 4-NM in the presence of NaBH$_4$. Reprinted from [34], Copyright (2020), with permission from Elsevier and [36], Copyright (2019), with permission from Elsevier as reprinted from [1], Copyright (2024), with permission from Elsevier.

better understand how effectively they function in increasingly complex real-world circumstances. This involves evaluating MNMs' performance in extensive waste-water/effluent management/treatment facilities and industrial settings. Additionally, it is critical to consider the possibility that MNMs could remove multiple contaminants at once, since this could greatly improve the efficacy of wastewater treatment and management procedures. In order to fully evaluate the ecological effects of MNMs, researchers should also concentrate on investigating their associated behavioral and environmental toxicity, creating suitable disposal plans, and putting LCAs into practice. Furthermore, since previous research has mostly focused on the potential for MNM recycling, it is imperative to investigate the stability and recyclability of these composites for up to about five cycles, particularly with regard to applications in commerce [35].

7.2.3 Applications of magnetic nanomaterials in catalysis

The immobilization of catalysts on support materials is an ongoing area of study in contemporary catalysis, with the goal of bridging the gap between heterogeneous and homogeneous catalysis. In view of their remarkable properties, which include high surface-to-volume ratios, low toxicity, increased thermal stability, increased catalytic action, as well as ease of dispersion and surface modification, MNMs are particularly notable in this field. Consequently, MNMs have become a frontrunner in the field of catalysis [37]. The current phase of Au catalysis is underway with the immobilization of Au nanocatalysts on MNMs. Several organic chemical reactions have been used to evaluate these nanostructured systems' catalytic effectiveness in depth [38]. In view of their small size, MNMs frequently form domain barriers that result in extremely energetic states and cause the particles to assemble or agglomerate. Numerous protective coatings, including silica, carbon, metallic, metal oxide, and polymeric layers, can be used to combat this tendency toward agglomeration. By encasing the MNMs, these coatings improve their stability and act as insulators against their external environment. By utilizing the substantial surface area that MNMs offer, a multitude of catalytically active substances can be arranged on their surface, consequently augmenting the activity of their catalysis [39].

The tendency of magnetic metal oxides, such as Fe_3O_4, $MnFe_2O_4$, and $CoFe_2O_4$, and metal alloys like $CoPt_3$ and FePt to magnetize in a hysteresis loop demonstrates their unique magnetic properties. Heterogeneous catalysis makes excellent use of these materials [40]. Additionally, a core–shell structure is formed by a variety of materials, including metals [41], metal oxides [42], silica [43, 44], polymers [45], and carbon [46, 47]. This enhances the functioning and stability of these materials. In order to catalyze the Suzuki–Miyaura reaction, Deng *et al* [48] for example, synthesized a catalyst using an N-heterocyclic carbene palladacycle immobilized on MNMs, which showed exceptional reactivity. MNMs are highly advantageous for hydrogenation (an important procedure in industrial catalysis and chemical synthesis) because of their large surface area that may accommodate active metal sites [49]. In a similar study, Yao *et al* [50] used Pd NMs immobilized on FeOx-coated Fe NMs to produce a Pd-based MNM for hydrogenating 2-methyl-3-buten-2-ol. Similarly, by adding CeO_2, Dinamarca *et al* [51] optimized the Fe_2O_3–SiO_2 structure and increased the selectivity of an Fe_2O_3–SiO_2–CeO_2–Pt catalyst in the conversion of cinnamaldehyde to cinnamic alcohol. MNMs also improve the purity and repeatability of catalysts in heterogeneous catalysis. A thorough investigation that used MMTPFe NM (heterogeneous catalytic MNMs or MNPs with immobilized montmorillonite (MMT)) to reduce 4-nitrophenol (4-NP) is shown in figure 7.2(b) by Das *et al* [36], as adapted and reproduced from Bushra *et al* [1]. In order to create an MNM–hydride complex, BH_4 must initially be adsorbed on MNMs. The simultaneous adsorption of 4-NP molecules occurs at a slower rate. These two procedures can be reversed. Subsequently, the decreased product (4-AP) desorbs from the NM surface due to the nitro groups' inhibition of hydrogen and electron transport within the MNM–hydride complex. This process enables the MMTPFe NM to regenerate its active surface for subsequent cycles.

Nevertheless, it is still challenging to fully comprehend the catalytic processes of MNMs. Combining density functional theory (DFT) calculations with experimental research may improve the creation of more economical, environmentally benign, and effective MNMs [1]. Hybrid MNMs are becoming more and more popular due to their capacity to enable material functionalization. These NMs provide significant surface area to immobilize active species, making them effective supports. Furthermore, they can be contained by beneficial magnetic nanocatalysts that are appropriate for use in industrial settings [1].

7.2.4 Applications of magnetic nanomaterials in the food industry

Food component analysis has a significant impact on regulatory structures and nutritional legislation, making it a global priority [1]. This calls for exact monitoring of intrinsic components and any adulterants, even at very low levels within intricate food composites. Hence, it is essential for scientific food studies to explore very sensitive modern analytical procedures [1]. In this regard, the use of different NMs in conjunction with NT/BNT presents a viable solution to the drawbacks of traditional extraction techniques. Of these NMs, MNMs have attracted a great deal of interest because of their remarkable chemical, optical, and magnetic characteristics, which include superparamagnetism, simple separation using magnets, and easy functionalization. Because of these characteristics, MNMs are a great option to improve the effectiveness and specificity of biomolecular interactions with complex composites. The ability of MNMs to be modified in both covalent and non-covalent forms, together with their affordability and ease of use, makes it easier to develop customized MNMs for certain analytical applications [1]. To improve the effectiveness of the extraction of organic molecules from food samples, a great deal of research has been done on the combination of MNMs and other carbon-based MNs [52]. Ulusoy et al [53], for example, used two different carbon-based NMs, multi-walled carbon nanotubes and nanodiamond, to create a MNM with Fe_3O_4. Its purpose was to extract vitamin B12 using high-performance liquid chromatography with a diode array detector (HPLC-DAD) from cereals and milk-based infant formula. Furthermore, Sereshti et al [54] used MNMs in combination with 3D graphene's special qualities to determine the amounts of aflatoxins in samples of bread. Additionally, to improve nutritional evaluation capabilities, metal–organic frameworks (MOFs), which are known for their large surface area and adjustable functionalization possibilities [55, 56], have been combined with MNMs [57, 58]. In order to extract aflatoxin B1 from herbal teas using magnetic solid-phase extraction (M-SPE), Durmus et al [59] developed a nanocomposite by coprecipitating MIL-53(Al) (Materials of Institut Lavoisier-53, an MOF) with core–shell $Fe_3O_4@SiO_2$ metallic NMs (m-NMs) ($Fe_3O_4@SiO_2@MIL$-53(Al)). In practical sample studies, this technique showed good analytical performance with a low limit of detection (LOD) of 0.50 ng ml^{-1}, a broad range of linear values from 0.50 to 150.00 ng ml^{-1}, and reasonable rates of recovery ranging from 70.70% to 96.50%. Similarly, Senosy et al [60] optimized important experimental variables to guarantee good sensitivity and reproducibility while developing a zeolite imidazolate framework-8 (ZIF-8)-enhanced magnetic graphene oxide NM for the extraction of triazole fungicides from diverse liquids. The resulting extraction method demonstrated a linear range of detection for all target analytes under optimum conditions, ranging from 1 to 1000.00 $\mu g\,l^{-1}$, with correlation coefficients (R^2) of $\geqslant 0.9914$. The four triazole fungicides had varying LODs and concentrations, ranging from 0.014 to 0.109 $\mu g\,l^{-1}$ and 0.047 to 0.365 $\mu g\,l^{-1}$, respectively. Furthermore, as studies [52, 53] have demonstrated, MNMs have been successfully used to analyze the inorganic and organic components in a variety of food substances, such as cereals, chocolate, drinks, eggs, fruits, honey, milk, and oils.

Regardless of the recent advances, there are many challenges that must be addressed in the food industry when using MNMs [1]. It is necessary to carry out an in-depth evaluation of the benefits and any possible risks linked to their utilization. Developing suitable legislation and implementing focused biosafety risk assessment methodologies are important yet difficult topics that need in-depth study. Novel testing approaches are essential for assessing MNMs' long-term effects on environmental and public health [61, 62]. Thorough investigation and monitoring are necessary due to concerns over the toxicity/noxiousness of NMs and their possible migration into food-related substances. One of the main goals is to guarantee that NMs remain stable throughout the entire sequence of food production, preservation, and distribution processes. More importantly, preserving safety and compliance depends on the advancement of reliable methodologies for the identification and characterization of MNMs. One of the biggest obstacles still remaining is addressing the issues of scalable and reasonably priced production techniques. It is also important to manage lingering doubts over public acceptability and attitudes toward NT/BNT in food products [63–67]. Furthermore, determining efficient waste disposal plans and evaluating the environmental impact of MNMs are essential for sustainable operations. To guarantee the safe and effective introduction of MNMs into the agricultural and food sectors, cooperation between scientists, regulators, and industry stakeholders is crucial. These collaborations will support the public's confidence in food products enhanced by NT/BNT while assisting in navigating the intricate regulatory framework [1].

7.3 The environmental life cycle analysis of magnetic nanomaterials

To identify both the beneficial and detrimental influences of NT/BNT advancements on the environment, it is essential to incorporate LCA into NT/BNT frameworks [1, 68–70]. This thorough analysis necessitates a comprehensive investigation of all aspects of the life cycle of a product, encompassing resource and raw material extraction, manufacturing procedures, product use, and end-of-life monitoring and management techniques such as recycling, reuse, and disposal. LCA assists in the early detection of any health or environmental hazards associated with the use of NMs by using this all-encompassing approach. However, there are a few challenges in applying LCA to NMs. Critical challenges arise from the lack of specific hazard-related data and characterization metrics for the assessment of environmental impacts within the LCA framework, an overall reluctance to reveal detailed information concerning current research and development procedures, and uncertainties arising from the recent development of NT/BNT and its volatile market dynamics [71]. Because of their intrinsic characteristics, wide range of uses, and disposal practices, MNMs present particular difficulties for the application of the analytical framework. Comprehensive knowledge of the environmental effects of MNMs during production, use, disposal, and recycling is essential, and the LCA approach can help with this. This methodology makes it possible to investigate how the synthetic routes used to produce MNMs affect the environment, including the energy required during the extraction of raw materials and the consumption of energy during development. In addition, LCA makes it possible to evaluate MNMs at every stage of their life cycle, including management, use, and final disposal. The application of LCA to MNMs, particularly multicomponent MNMs, is still restricted to an insignificant number of cases, even though substantial studies have been conducted using the consumer product assessment framework [1]. In view of this, this chapter primarily relies on research into single-component MNMs, while acknowledging that findings from these studies are frequently applicable to multicomponent MNMs.

The ISO 14040-44 guidelines describe the four steps of the LCA technique that are used to assess the environmental impacts related to MNM syntheses [72]. These steps are:

Goal and scope definition (step 1): to establish the scope of the assessment, this step entails demarcating system boundaries and identifying functional units (FUs) for various production sizes.

The life cycle inventory (LCI) (step 2): to complete this step, all pertinent data for every stage of the synthetic process should be systematically gathered. This entails determining the fluxes of energy, water, raw materials, and waste/effluent which are necessary for the manufacture of MNMs.

The life cycle impact assessment (LCIA) (step 3): considering their possible influence on the environment, the gathered inventory data are classified in this step. This categorization aids in comprehending the ways in which different procedural inputs and outputs impact particular environmental areas of study.

The interpretation (step 4): in this final step, the data are analyzed to determine the main causes of environmental loads, identify important environmental risk areas, conduct sensitivity studies, and consider other options. This aids in reducing adverse effects on the environment and improving the overall performance of the MNM synthesis procedure with respect to the environment.

Everything that occurs, including various processes and activities that contribute to the life cycle of NMs, is identified by the system boundaries in LCA [73]. Historically, the cradle-to-gate (from raw materials to manufacture) method has been the main focus in LCA studies of MNMs. On the other hand, taking a cradle-to-grave viewpoint (covering the full life cycle) might provide a more thorough comprehension of the environmental effects connected to these materials. A key idea in LCA is the FU, which is the measured performance of a product system that may be used as a benchmark. When referring to MNMs, the FU is determined by the unique performance attributes of the NMs, such as mass or specific application features. Following that, the effects on the environment are quantified and contrasted across a few different categories of impact using this FU. Thoroughly identifying data uncertainties and accounting for margins of uncertainty are also necessary [74].

Since the production of MNMs usually uses hazardous raw chemicals, it is imperative to replace these compounds with more environmentally friendly reagents, stabilizing agents, and reducing agents. A shovel-to-gate LCA was used to assess the environmental effects of several MNM production techniques. The main goal of this study by Feijoo *et al* [75] was to replace hazardous raw materials in the production processes with more environmentally friendly alternatives. They synthesized and examined a variety of MNMs coated with NMs such as silica, oleic acid, sterically stabilized Fe_3O_4, and polyethylenimine (PEI). Their research found that reagents utilized for the adjustment of pH, such as HCl, were a key source of environmental consequences and that particular compounds, such as $FeCl_3$, NH_4OH, and TMAOH, used in the synthesis/production procedure, contributed significantly to environmental burdens. These findings suggest that less harmful alternatives are needed. During the manufacturing process, energy and chemical usage were found to be the main elements influencing the environmental implications, especially for silicon-coated NMs. PEI-coated NMs are a more favorable choice because of their balanced performance and lower environmental effect than silica-coated NMs, which were found to have major environmental disadvantages despite their greater performance. In a similar vein, Baresel *et al* [76] studied MNMs utilized during water treatment, investigating the relationship between their LCA and their recovery rate. After synthesizing MNMs intended to absorb HM ions, they found that the global warming potential (GWP) of these substances was significantly decreased by their higher recovery rate. The study revealed that the primary source of environmental consequences was raw materials, underscoring the significance of the recovery rate in mitigating the environmental impact of MNMs. The FU was defined as the dosage of an MRI contrast agent for a patient weighing 70.00 kg. Their analysis was based on LCA data collected from two pilot-scale plants and previous studies. They lowered resource consumption and ozone depletion by eliminating the usage of hazardous chemicals by improving their synthesis process. Their strategy also reduced workplace risks and consumed less energy. In general, their research indicates that, compared to wet chemical synthesis, laser vaporization is a more sustainable way to produce MNMs.

Upon comparing large-scale coprecipitation with other synthetic procedures in terms of environmental impact, it becomes clear that this process is preferable for producing Fe_3O_4 NMs due to its eco-friendliness. Compared to traditional coprecipitation techniques, the use of glutathione as a reductant and stabilizer during the course of production/synthesis significantly lessens its environmental impact [77]. The production/synthesis procedure shows a considerable decrease in the consumption of energy, resulting in a tenfold reduction in GWP while maintaining equivalent levels of toxicity to humans, even when comparable quantities of raw materials are employed. In this 'green synthesis,' sodium hydroxide (NaOH) and glutathione are recognized as significant factors that affect the environment. Specifically, glutathione reduces the requirement for extra energy in material-intensive processes by facilitating the synthesis of –SH groups on its reactive surface, which lowers the demand for additional coating stages. Bushra *et al* [1] have presented an additional list of other LCA studies that evaluate several environmental issues, including GWP, environmental toxicity, and resource depletion, providing a more comprehensive understanding of the environmental effects of MNMs.

The majority of research shows that throughout the life cycles of items and procedures, the main factors influencing environmental impacts are energy and chemical use. Notably, up to 60% of these adverse environmental effects are attributable to the use of chemicals [75]. As a result of these findings, there is an immense drive to substitute environmentally sustainable raw materials for conventional ones when producing MNMs chemically. One practical way to upcycle and lessen the environmental impact associated with NM manufacture is to use waste as a feedstock for the development and advancement of NMs/NPs [3, 5, 78, 79]. Joshi *et al*'s [80] study of the microbial reduction of naturally occurring Fe(III) to Fe(II) NMs shows that, in comparison to traditional approaches, biosynthetic synthesis dramatically lowers costs and environmental implications. Their LCA indicated declines of 3.53 MJ in natural gas use, 13.73 MJ in the utilization of fossil resources, 4.97 m^3 for water usage, and 0.77 kg CO_2 equivalent in GWP. Due to these efficiency improvements, the cost of NMs was reduced by about 315 € kg^{-1}, demonstrating the significant potential of biosynthetic techniques to significantly reduce resource usage and environmental impact during NM production. Although the life cycle analysis of NMs is becoming more widely acknowledged in academic and industrial domains, there is still a significant lack of dependable and comparative research in this field. This discrepancy causes substantial uncertainty when estimating the environmental effects of their use. The abovementioned main obstacles include the early stages of magnetic nanomaterial (MNM)/magnetic nanomaterials (MNMs) technologies and their growth in the market, an apparent unwillingness to reveal specific synthetic procedures, and a lack of knowledge regarding the end-of-life management and breakdown processes of MNMs. Further studies are anticipated to shed additional light on this area and broaden our comprehension. Data on the environmental impacts of MNM production should be made accessible to everyone in order to help shape future sustainable practices. MNMs show promise for a wide range of applications, although this field of study is still in its infancy. Currently, three main obstacles are impeding the development of MNMs: the technology for NM synthesis is still in its infancy; there is a dearth of detailed information regarding necessary experimental conditions; and there are difficulties in defining the FU, which is necessary to standardize assessments of impact. Moreover, circularity principles ought to be aligned with the quest for sustainability in MNM engineering. In this context, utilizing MNMs to extend a product's life cycle is essential to advancing circular economy principles. When MNMs are used in catalytic processes, for example, waste can be reduced since they can be repeatedly reused and are easier to recover using external MFs [81].

When considered collectively, the few research projects in this area utilizing the LCA approach have determined that energy use and raw material utilization account for the majority of the environmental effects. As a result, it is advisable to use alternative raw materials to enable cleaner MNM production processes. It is also crucial to remember that NMs can be fully recovered using external magnetic fields, which can drastically lessen their environmental impact—possibly by up to 835 times. This discovery creates opportunities to investigate MNMs with reduced ecological impacts [1].

7.4 Hazards and risk management associated with magnetic nanomaterials

Due to their numerous uses in fields ranging from materials science to biological processes, MNMs have attracted a lot of attention. However, just like other technologies, there are dangers and hazards associated with using MNMs. Ensuring the safe integration of these hazards requires a full understanding of them and competent risk management [1]. To support this discussion, this section examines the risks connected to MNMs and expounds upon the core ideas of risk assessment and mitigation.

7.4.1 Hazards associated with magnetic nanomaterials

In general, MNMs and other NMs are more hazardous than their bigger counter-parts formed of the same materials [1, 82]. The capacity of NMs to penetrate deeply into bodily structures, particularly the cell nucleus, which is impenetrable to larger particles, is the reason for this heightened toxicity. This penetrating tendency could lead to a buildup in human cells, which may be harmful [1, 82]. Although NMs are usually categorized in risk evaluations and ecotoxicology, there are still issues because there is a dearth of detailed information [83], especially with regard to their possible toxicity. Due to their tiny dimensions and large surface area, MNMs interact differently with biological structures, which raises issues. The amount of toxicity varies according to MNMs' size, shape, surface coating, and surface charge, among other parameters. Research has indicated that particular MNMs can cause biological entities to experience oxidative stress, genotoxicity, or cytotoxicity [84]. In light of their ability to decompose and produce ferric ions (Fe^{3+}), MNMs were initially thought to be nontoxic; however, further research has indicated that they may become hazardous when they accumulate in cells [85]. Polymer coatings, such as poly(ethylene glycol) (PEG) and polyglycerol, are frequently added to MNMs in order to mitigate their possible toxic effects. These coatings work by reducing direct cellular contact, improving hydrophilic qualities, preventing agglomeration, and reducing toxic effects [86]. For example, in comparison to uncoated MNMs, surface coatings reduce toxicity by increasing hydrodynamic diameter. Furthermore, because surface-charged NMs interact with and are absorbed by cells more readily than neutral ones, they typically exhibit higher degrees of toxicity. Hafiz *et al* [87] presented an overview of the toxicological effects related to MNMs, including their coatings, by aggregating data from various studies. Uncoated MNMs have been found to cause physiological alterations in living creatures at concentrations higher than 10 mg l^{-1} [88], and at concentrations of 25 mg l^{-1}, red blood cell destruction has been noted [88, 89]. On the other hand, compared to their uncoated counterparts, coated MNMs, such as chitosan-coated MNMs, show reduced toxic effects [90]. However, chitosan-coated MNMs' higher surface charge increased their intracellular uptake, raising further safety questions [91]. Due to the intrinsic low toxicity and biocompatibility of polyethylene glycol polymers, PEG-coated MNMs showed no adverse effects even at concentrations as high as 100.00 mg l^{-1} [92]. In a comparable manner, when used at a significant dose of 200 mg l^{-1}, polyglycerol-coated MNMs showed no toxic effects [93].

With regard to MNMs' exceptional biological compatibility in BM applications, careful assessments of their influence on immune system responses, tissue compatibility, and possible long-term consequences within the human system are required. It is essential to comprehend the intricate relationships between MNMs and biological systems in order to limit negative consequences. Moreover, MNMs have the ability to penetrate the environment via a variety of routes, endangering both land and water ecosystems as well as possibly impacting microbes, plants, and animals. Because MNMs can enter an individual's body by ingestion, inhalation, or skin contact during production and study operations, occupational exposure raises additional issues. For risk evaluations to be efficient, an in-depth knowledge of these exposure routes and the quantification of exposure levels is essential. Furthermore, MNMs' powerful magnetic fields have the ability to interact with magnetic materials found in the human body's tissues. In certain situations, this interaction may cause mechanical strain or interfere with electronic equipment such as cardiac pacemakers, which could pose a risk to patient safety. Guidelines to reduce the dangers of compounds such as MNMs may need to be codified into law depending on a number of variables, particularly the product's nature and quantity. The Registration, Evaluation, Authorisation, and Restriction of Chemicals (REACH) guidelines of the European Community do not specifically address MNMs, but it is suggested that because MNMs are classified as a 'substance' under the guidelines they establish, these guidelines cover the safe and standardized application of NMs by default. The Scientific Committee on Emerging and Newly Identified Health Risks (SCENIHR), on the other hand, emphasizes the need for reliable assessment techniques and instruments for NMs, pointing out their diverse toxicological characteristics as well as the requirement for case-by-case analyses. Consequently, there is a pressing necessity for an in-depth investigation into the toxic oral effects of MNMs, and legislative bodies should give this issue top priority [1].

7.4.2 Risk assessment associated with magnetic nanomaterials

To ensure the safe application of MNMs, risk evaluation is a crucial and thorough procedure. The first step in the procedure is the identification of hazards, which requires an extensive understanding of the physical, chemical, and biological characteristics of MNMs in order to foresee any potential adverse consequences. Exposure evaluation, which critically examines the possibilities of contact between MNMs and the surrounding environment or humans, comes after this stage. As part of this investigation, exposure routes, exposure concentrations, and interaction durations are assessed. Four essential components make up the traditional risk assessment structure, which was initially developed in the US and is still used in many countries [94]: the identification of hazards, exposure assessment, dose–effect relationship assessment, and risk characterization. By establishing the relationship between MNM levels of exposure and the likelihood and consequences of potential effects, dose–response evaluation allows for the assessment of risk in many contexts. After that, risk characterization aggregates all the information gathered to evaluate the total risk while taking into account the probability and seriousness of any harm. Thoroughly recognizing and evaluating uncertainty and properly conveying the results to relevant stakeholders are prerequisites for efficient risk management. Utilizing personal protective equipment, encouraging safe work habits, establishing engineering controls, and ensuring that workplaces have adequate ventilation are a few examples of risk mitigation techniques. Defining established processes and ensuring that they are followed in order to handle, store, dispose of, label, and contain MNMs is also essential. Comprehensive testing for biocompatibility is essential in applications in BM sciences. Furthermore, in order to reduce adverse impacts on the environment, adequate storage and treatment of wastewater are essential steps in preventing MNM discharge. In addition, the implementation of monitoring systems, emergency response plans, and thorough training and education are essential components of a strong risk mitigation approach, as are ethical research procedures. Efficient risk mitigation requires ongoing risk evaluation, modifying safety procedures with emerging information, clearly communicating hazards, and investigating substitute materials or tactics. Research into substitute materials or methods to reduce risks without sacrificing effectiveness is a promising direction [95, 96].

In short, a comprehensive and adaptive procedure is required to manage the risks related to MNMs. This covers risk estimation, assessment of exposure, identification of hazards, and the implementation of mitigation strategies. The main objective is to guarantee that MNMs are utilized responsibly and securely in a variety of disciplines, with a methodology that adjusts to new discoveries and perceptions [1].

7.5 Market variability, prospects (opportunities), and economic sustainability of magnetic nanomaterials

The anticipated size of the worldwide market for MNMs in 2022 amounted to 74.20 million US dollars. This is anticipated to increase to 139.20 million US dollars by 2028, indicating a compound annual growth rate (CAGR) of 10.40% from 2023 to 2028, as reported by IMARC Group, a reputable market research firm. MNMs are essential for several applications, such as targeted delivery systems for cancer therapy, medication nanocarriers, and diagnostics, where they act as contrast agents and nanoprobes. The MRI industry is anticipated to grow at a CAGR of more than 5.00% from 2017 to 2027, when it is expected to reach a potential market size of eight billion US dollars [1, 97, 98]. Within a comparable time frame, the MNM market is anticipated to have substantial potential due to its current growth trajectory. Furthermore, considering that MNMs are used in the identification and management of diseases, including cancer and brain tumors, the market for MNMs is anticipated to benefit from the increasing number of chronic and life-threatening illnesses. Given that MNMs are used in magnetic bioseparation, superparamagnetic relaxometry, and the identification of biological entities such as bacteria, viruses, proteins, nucleic acids, and cells, there is also anticipated to be an increase in demand for MNMs. Notwithstanding these encouraging developments, research is still being conducted on issues including NM toxicity, scalability, and cost-effectiveness. The market is anticipated to grow further as a result of ongoing scientific developments and growing applications in fields such as BM, electronics, environmentally friendly products, and other areas as different businesses continue to find novel applications and capitalize on the special qualities of MNMs. Iron-based and cobalt-based MNMs make up the two main categories of the global MNM market, with iron-based MNMs presently possessing the highest market share [1]. The global marketplace offers three different physical forms: dispersion, solution, and nanopowder. Of these, nanopowder is the most commonly used. MNMs are used in a wide range of industries, including BM, electronics, wastewater/effluent monitoring and management, energy, etc. The BM industry holds the highest market share. The market is divided geographically into the following regions: North America, Latin America, Europe, Asia-Pacific, the Middle East, and Africa, with North America occupying the top spot worldwide. According to Bushra *et al* [1], the following enterprises are prominent players in the MNM market: Reade International Corp., SkySpring Nanomaterials Inc., Strem Chemicals Inc., American Elements, Cytodiagnostics Inc., Merck KGaA, Nano Research Elements Inc., nanoComposix (Fortis Life Sciences), Nanografi Nano Technology, Nanoshel LLC, and US Research Nanomaterials Inc.

7.5.1 Market variability of magnetic nanomaterials

Growth is hampered by security challenges, laws, and regulations in a number of industries, particularly biotechnology, healthcare, and innovative technologies. Although adhering to stringent laws is essential for upholding safety, effectiveness, and ethical standards, doing so can be expensive and time-consuming. Emerging technologies (for example, artificial intelligence (AI) and driverless cars) face enormous obstacles related to safety certification, liability concerns, and ethical considerations. These issues may cause public discussion and regulatory scrutiny to increase, which would impede the commercial development of such technologies. Given that they are more expensive for end users, the use of specialized materials and innovative manufacturing methods could restrict their general implementation. Complex assembly and stringent quality control protocols drive up costs in industries such as electronics, renewable energy, and BM equipment. Similarly, the biopharmaceutical industry requires strict compliance with regulatory requirements and substantial research via clinical trials in order to produce new treatments, particularly biologics. Patients and healthcare providers bear greater financial costs as a consequence of these essential but highly resource-intensive procedures [1].

7.5.2 Prospects (opportunities) of magnetic nanomaterials

7.5.2.1 Rising desire for biomedical applications

There are several opportunities to enhance patient outcomes and develop healthcare due to the increasing need for BM technology [1]. Numerous fields are covered by these applications, including tissue engineering, medicine, medicine delivery, imaging, treatments, and diagnostics [18, 21, 22]. Persistent technical developments, such as the identification of biomarkers and molecular imaging, make it possible to diagnose illnesses earlier and more precisely. By stopping the spread of diseases, this in turn makes it easier to implement interventions in a timely manner and may result in cost savings. Additionally, the creation of customized medicines is being fueled by the combination of NT/BNT [1]. Targeted medicine delivery with NMs reduces negative effects and improves treatment effectiveness. Moreover, the fusion of state-of-the-art NMs with three-dimensional printing technology facilitates regenerative medicine by permitting the synthesis of functioning tissues and organs [1].

7.5.2.2 Growing interest in nanotechnology/bionanotechnology

NT/BNT is becoming more and more popular owing to its ability to reduce the size of electronic components, which increases the speed and energy efficiency of devices. Ultrahigh-resolution screens, flexible electronics, and powerful computers are all products of nanoscale inventions such as transistors, quantum dots, and nanowires. NT/BNT is transforming targeted drug delivery in the BM field, increasing therapeutic efficacy while reducing negative effects. The visualization of biological structures is greatly enhanced by nanoscale imaging agents, which improve early illness identification and accurate therapy monitoring. Furthermore, by using NMs for water purification, pollution remediation, and enhanced energy storage in solar cells, batteries, and supercapacitors, NT/BNT also addresses environmental problems. There are many opportunities due to the versatile applications and revolutionary potential of MNMs in several scientific and industrial sectors [1]. It is anticipated that future study and development will uncover novel applications, confirming the significant influence of MNMs on technological development and research in science.

7.5.3 Economic sustainability of magnetic nanomaterials

Owing to their distinct nanoscale characteristics, MNMs have notable economic viability and hold promise for a wide range of applications in industry. For stakeholders as well as investors looking to capitalize on the available opportunities, understanding their economic viability is crucial. MNMs are being investigated in the BM field for targeted medication delivery, diagnostics, and imaging applications. Although MNM-based medicines and diagnostic tools have a high upfront cost, their ability to precisely administer pharmaceuticals to specific cells or tissues can enhance treatment outcomes, reduce side effects, and save long-term costs for patients and the healthcare system [1]. MNMs also contribute to the advancement of electronics, particularly sensors, data storage, and cutting-edge fields like spintronics. The development and integration of MNM-based technologies may prove costly at first, but over time, improvements in device effectiveness, energy efficiency, and longevity can make electronic devices more durable and cost-effective. MNMs are essential to environmental management because they effectively remove contaminants from soil and water, providing environmentally acceptable solutions. They also play a significant role in pollution management and monitoring, as well as in water purification. MNM-based technologies have the potential to lower healthcare costs and create cleaner ecosystems, which can lead to overall economic advantages even if their implementation may require large upfront investments. It is critical to recognize that different applications, manufacturing volumes, and regulatory considerations may impact the economic viability of MNMs. Innovation is anticipated to be sparked by ongoing studies and developments, which may reduce manufacturing expenses and increase the viability of MNMs across a variety of industries [1].

In summary, MNMs have significant economic potential in industries including electronics, environmental management, and BM, as well as in the NT/BNT sectors. As the range of applications for MNMs increases, investment in their research and development is expected to yield substantial and long-lasting benefits that will encourage economical solutions and long-term economic sustainability [1].

7.6 Techniques for the recovery and recycling of magnetic nanomaterials

For MNMs to be used in a variety of applications, their recovery and recycling capabilities are crucial. MNMs' inherent magnetic characteristics provide effective processes for separation and retrieval, which accelerates their recovery after application [16, 99, 100]. Due to their innate capacity for recycling, they promote resource efficiency and reduce harmful impacts on the environment, in line with sustainability standards. However, a number of issues, such as agglomeration tendencies and restricted recyclability in some extreme circumstances, exist, necessitating careful thought for effective application [1]. These issues are the subject of ongoing research, which attempts to improve the general sustainability and usefulness of MNMs. For MNMs to be used in scientific and industrial fields in a sustainable and efficient manner, a thorough examination of their recovery and capacity for recycling is essential [1].

7.6.1 Recovery techniques for magnetic nanomaterials

When MNMs have served the purpose they were designed for, they need to be isolated and extracted from a composite or solution. This procedure is known as MNM recovery. To achieve effective MNM recovery, many approaches are used. These techniques include filtration, centrifugation, and magnetic separation. Other, more specialized techniques encompass chemical precipitation, gradient MFs, ultracentrifugation, magnetic fluid heating, and recovery assisted by electromagnetic fields (EMFs). As highlighted by Bushra *et al* [1], the following sections summarize a few selected techniques.

7.6.1.1 Magnetic separation technique

One method that is frequently used to collect or isolate MNMs is magnetic separation. In this technique, an MF is applied to attract MNMs using permanent magnets, electromagnets, or specialized magnetic separators. To efficiently collect and remove the NMs from the solution, the MF's strength and arrangement can be changed. In the field of catalysis research, magnetic separation is especially note-worthy [101, 102]. By isolating and extracting the desired product, it simplifies post-reaction processes. This process eliminates the need for traditional separation methods such as centrifugation and filtration. Most significantly, it facilitates complete product isolation without the need for solvents, which is helpful in batch and laboratory procedures where the inclusion of extra solvents is frequently required. However, a strong metal-supporting substrate is necessary for the separation procedure to be effective, and this interaction can be enhanced by functionalizing the support surface [103]. One method that is frequently used to gather or isolate magnetic substances from flowing streams is the separation of magnetic materials [1].

7.6.1.2 Gradient magnetic field technique

MNMs are recovered and separated according to their size and magnetic suscept-ibility using gradient MF. The process described here uses a technique called high gradient magnetic separation (HGMS) to attract the particles [104]. This technique is often used in many different magnetic separation procedures and is especially useful for separating magnetic entities from a nonmagnetic medium. Larger particles or conglomerates have historically been separated at the micron scale using HGMS [105]. Earlier research examined its effectiveness in removing Fe_3O_4 nanoclusters and iron oxide nanoparticles (IONPs) from aqueous solutions [106, 107]. The effectiveness of magnetically charged particle collection depends on the magnetic force's ability to overcome the forces of gravity, inertia, fluid drag, and diffusion during the passage of the particle suspension through the separation system [1].

7.6.1.3 External-magnetic-field-assisted recovery technique

The EMF-assisted recovery (EFAR) technique has a number of positive aspects for MNM recovery, particularly its non-invasiveness, scalability, and capacity to target certain NMs. This method is especially useful in a variety of domains, including cleaning up the environment and therapeutic applications such as menopausal hormone therapy (MHT)-assisted cancer treatment and targeted drug delivery. Depending on the features of the MNMs and the application's aims, different parameters, tools, and operating conditions may be employed in EFAR [1]. Scientists and engineers frequently modify parameters to improve the recovery process and customize it to meet their own requirements. Research has demon-strated the effectiveness of the technique; for example, a study on the effect of EMF on the recovery of oil factors revealed a considerable increase. A 35.45% recovery rate was achieved using IONPs in the absence of an EMF. On the other hand, the oil recovery rate increased with the introduction of an EMF, showing an impressive 79.83% rise [108]. Furthermore, it has been demonstrated that the use of tailored

MNMs in dynamic EMF-assisted solid-phase extraction is an effective and quick analytical method for removing flavonoids from onion samples [109].

7.6.1.4 Centrifugation technique

An additional technique for obtaining NMs is centrifugation, which is particularly beneficial when working with greater quantities or complicated mixes. By quickly spinning the combination at extremely high speeds, this approach makes it possible to separate MNMs from magnetically inactive components according to the manner in which their densities vary. Scientific and technological laboratories frequently use this approach in research. To separate superparamagnetic iron oxide nanoparticles (SPIONPs) with accurately determined dimensions and very narrow size distributions, for example, Dadfar *et al* [110] highlighted a stepwise centrifugation technique. With core sizes of 7.7 ± 1.6, 10.6 ± 1.8, 13.1 ± 2.2, 15.6 ± 2.8, and 17.2 ± 2.1 nm, they obtained hydrodynamic diameters of 26.3 ± 1.2, 49.4 ± 1.1, 64.8 ± 2.1, 82.1 ± 2.3, and 114.6 ± 4.4 nm. This resulted in a polydispersity index (PDI) of less than 0.1 from an originally polydisperse mixture of SPIONPs.

7.6.2 Recycling techniques for magnetic nanomaterials

MNMs' capacity to be recycled is essential for increasing cost-effectiveness and sustainability in a variety of applications. MNMs frequently have excellent potential for regeneration and multi-cycle utilization, which lowers the requirement for acquiring fresh NMs. In catalytic applications, for instance, MNMs are readily retrieved after the reaction, cleaned, and reintroduced into further reactions. This strategy lowers the expenses related to the use of MNMs while also minimizing waste output. By managing surface qualities and resolving problems such as agglomeration, stability, compatibility, and the prevention of environmental impacts, surface changes play a crucial role in improving the recyclability of MNMs. These changes increase MNMs' capacity for recycling and enhance their usefulness across a variety of industries. To create reusable MNMs, several methods have been proposed [111, 112]. The financial and practical usefulness of surface-functionalized MNMs for applications in industry is further highlighted by their regenerative and reuse capabilities, which go beyond their outstanding properties and high effectiveness. In summary, one of the most important aspects of MNM applications is their reusability. For the recovery and subsequent reuse of these NMs, a variety of techniques, such as precipitation by chemicals and separation by MF, can be used, enhancing sustainability and minimizing environmental effects. It is essential to customize the recycling process to the unique properties of the MNMs and the intended use in order to attain the best possible outcomes [1].

7.7 Utilization difficulties and commercial challenges of magnetic nanomaterials

Applications for MNMs have increased significantly in a variety of fields, including electronics, BM, and the environment. However, they are also subject to limitations, as highlighted by Bushra *et al* [1]:

(a) Both the intrinsic magnetic properties and biophysical properties of MNMs determine the best modification for biological applications. Surface chemistry must be carefully considered in order to avoid MNM accumulation, which calls for stabilization using a variety of materials, including inorganic and organic protective coatings [113]. It is difficult to create surface functionalization techniques that work well, since they have a significant impact on MNMs' stability, reactivity, and possible toxic effects.

(b) Considering that MNMs can potentially be sensitive to variations in pH, temperature, or ionic strength, it can be difficult to keep them stable in a variety of settings. MNMs' specific requirements and limitations vary depending on whether they are utilized *in vivo* or *in vitro*. To avoid toxic effects, reduce the chance of blood capillary blockage, and stop NM aggregation, it is essential to coat or cover magnetic structures intended for *in vivo* applications with a biodegradable substance before or after manufacture.

(c) The inherent constraints of *in vivo* investigations have posed challenges for the therapeutic translation of MNM-mediated hyperthermia. These constraints are mostly caused by the imprecision and decreased heat production that occur when cells endocytose MNMs. As a result, research into various morphologies, structures, and designs has focused on improving the specific absorption rate of MNMs. Enhancing the efficiency of nanoscale thermal therapeutic applications seems to most likely involve optimizing the nanoscale transfer of energy and attaining targeted intracellular localization at the tumor location while reducing the nonspecific deposition of MNMs in other parts of the body.

(d) MNM-based drug release techniques are currently limited by their inability to penetrate tissues well and their potential for cytotoxicity. Progress in the future depends on resolving these problems. The goal of future studies should be to develop sophisticated, 'smart' MNMs that can release genetic material and medications in a regulated, sequential manner at precise times and places. Designing tailored treatments with ideal profiles can be enabled by taking advantage of tumor-specific microenvironment triggers, like acidic pH, to control drug release. This progresses the incorporation of MNMs into personalized medicine [114, 115].

(e) A thorough understanding of the toxicological aspects associated with NMs and nanosystems, as explored through both *in vivo* and *in vitro* investigations, is crucial for the future development of MNM-based BM science. While functionalized MNMs are widely used in therapeutic applications, it is also crucial to evaluate their long-term stability [116].

(f) An ongoing major difficulty is precisely regulating the morphology, size, and chemical structure of MNMs. Changes in these properties could have a significant impact on their magnetic properties and usefulness in particular applications.

(g) The production of MNMs frequently necessitates elaborate and expensive processes, which pose obstacles to their widespread commercial implementation. Moreover, the dimensional, morphological, and physicochemical properties of these NMs are strongly influenced by the synthetic methodologies chosen.

(h) To reduce the hazards resulting from long-term accumulation, it is imperative to guarantee the efficient removal of MNMs from biological structures for *in vivo* applications. An in-depth knowledge of the MNM clearance processes and all possible toxicological effects is a crucial component of this procedure. Moreover, regulatory permission is usually needed before MNMs may be used in medical applications, which is an expensive and time-consuming process.

(i) Hysteresis affects MNMs, causing a delay in their magnetic reaction to variations in external MFs. This property might impose limitations in situations where quick magnetic reactions are essential.

(j) Complex procedures are frequently used in the manufacture of MNMs, which may call for the utilization of potentially hazardous substances or high temperatures. The technical intricacies involved present significant obstacles to the scale-up of production processes.

(k) The cost of producing superior MNMs can be high, which may prevent them from being widely used in some applications.

(l) MNMs have great potential for a variety of BM applications; however, it usually takes longer for them to transition from the research lab to actual clinical applications. This is mostly because extensive testing and verification are required to ensure the products' safety and effectiveness.

(m) Meticulous consideration and mitigation are necessary for the disposal consequences and possible environmental effects linked to MNMs, especially those that comprise toxic elements.

(n) It can be difficult to obtain rights to intellectual property for novel MNM technology because of the intricacy of the current patent system.

(o) Different applications have different requirements and obstacles, such as MHT for cancer treatment, targeted drug administration, and the use of MNMs as MRI contrast agents. The process of customizing MNMs to fulfill the demands of many different uses may prove intricate and multifaceted.

Given the current limitations, studies are currently focussing on a number of these problems, enabling MNMs to continue proving their potential in a variety of contexts. Specific restrictions are mitigated in concert by advances in materials science, the field of engineering, modification of surface techniques, and fine control over particle size and morphology [1]. The cumulative progress made in these domains is expanding the range of applications for MNMs, highlighting their adaptability and potential influence in other fields of science and technology as well as in the NT/BNT domains.

7.8 Conclusions and future perceptions of magnetic nanomaterials

As a result of their special qualities, MNMs have great potential in a variety of fields, including environmental management and BM activities as well as in the NT/BNT domains. However, challenges to the environment and public health, as well as operational and economic limitations, make their industrial implementation challenging. In order to address these issues, this chapter suggests better environmental tactics, safety enhancements, and MNM synthesis advances. This chapter emphasizes the significance of toxicological evaluations and LCA in guaranteeing the safe and sustainable use of MNMs. To overcome these challenges and realize the full economic potential of MNMs, research and strategic developments are essential. MNMs are also becoming recognized as cutting-edge therapeutic instruments that could revolutionize cancer detection and therapy. MNMs are created using a variety of synthesis techniques, such as coprecipitation, thermal degradation, hydrothermal procedures, and polyol synthesis. The use of organic as well as inorganic polymers for functionalization has greatly improved MNMs' biocompatibility. Functionalization clinical studies show that MNMs have a great deal of potential, but long-term toxicological evaluations still lack sufficient information, and the ecological effects of MNMs are frequently overlooked, which is essential for increasing MNMs' effectiveness in BM applications such as medication delivery, mental health therapy, and diagnostics. While several environmental and clinical trials have shown MNMs to have great potential, there is still a clear lack of long-term toxicological evaluations, and the environmental consequences of MNMs have frequently been overlooked. In order to ensure their safe and efficient use in environmental and BM applications and advance the NT/BNT domains, it is essential to address these elements and obtain a thorough grasp of their implications.

References

[1] Bushra R, Ahmad M, Alam K, Farzad Seidi F, Qurtulen, Shakeel S, Song J, Jin Y and Huining Xiao H 2024 Recent advances in magnetic nanoparticles: key applications, environmental insights, and future strategies *Sustain. Mater. Technol.* **40** e00985

[2] Zhang K, Song X, Liu M, Chen M, Li J and Han J 2023 Review on the use of magnetic nanoparticles in the detection of environmental pollutants *Water* **15** 3077

[3] Ukhurebor K and Aigbe U 2024 *Environmental Applications of Magnetic Sorbents* (Bristol: Institute of Physics Publishing)

[4] Ukhurebor K, Aigbe U and Onyancha R 2023 *Adsorption Applications for Environmental Sustainability* (Bristol: Institute of Physics Publishing)

[5] Aigbe U, Ukhurebor K and Onyancha R 2023 *Magnetic Nanomaterials: Synthesis, Characterization and Applications* (Berlin: Springer Nature)

[6] Singh R, Ukhurebor K, Singh J, Adetunji C and Singh K 2022 *Nanobiosensors for Environmental Monitoring—Fundamentals and Application* (Berlin: Springer Nature)

[7] Yang Q, Dong Y, Qiu Y, Yang X, Cao H and Wu Y 2020 Design of functional magnetic nanocomposites for bioseparation *Colloids Surf. B: Biointerfaces* **191** 111014

[8] Van de Walle A *et al* 2020 Magnetic nanoparticles in regenerative medicine: what of their fate in stem cells? *Mater. Today Nano* **11** 100084

[9] Black C, Gates S, Murray C and S S 2000 Magnetic storage medium formed of nanoparticles *USA Patent US6162532A*

[10] Mohtasebzadeh A, Ye L and Crawford T 2015 Magnetic nanoparticle arrays self-assembled on perpendicular magnetic recording media *Int. J. Mol. Sci.* **16** 19769–79

[11] Sun S, Murray C, Weller D, Folks L and Moser A 2000 Monodisperse FePt nanoparticles and ferromagnetic FePt nanocrystal Superlattices *Science* **287** 1989–92

[12] Reiss G and Hütten A Applications beyond data storage *Nat. Mater.* **4** 725–6

[13] Skumryev V, Stoyanov S, Zhang Y, Hadjipanayis G, Givord D and Nogués J 2003 Beating the superparamagnetic limit with exchange bias *Nature* **423** 850–3

[14] Phan M-H, Alonso J, Khurshid H, Lampen-Kelley P, Chandra S, Stojak Repa K, Nemati Z, Das R, Iglesias O and Srikanth H 2016 Exchange bias effects in iron oxide-based nanoparticle systems *Nanomaterials* **6** 221

[15] Döpke C, Grothe T, Steblinski P, Klöcker M, Sabantina L, Kosmalska D, Blachowicz T and Ehrmann A 2019 Magnetic nanofiber mats for data storage and transfer *Nanomaterials* **92** 9

[16] Aigbe U, Ukhurebor K, Onyancha R, Osibote O, Kusuma H and Darmokoeso H 2022 Measuring the velocity profile of spinning particles and its impact on Cr(VI) sequestration *Chem. Eng. Process.* **178** 1–15

[17] Aigbe U, Ukhurebor K, Onyancha R, Okundaye B, Pal K, Osibote O, Esiekpe E, Kusuma H and Darmokoesoemo H 2022 A facile review on the sorption of heavy metals and dyes using bionanocomposites *Adsorp. Sci. Technol.* **2022** 8030175

[18] Ukhurebor K, Onyancha R, Aigbe U, UK-Eghonghon G, Kerry R, Kusuma H, Darmokoesoemo H, Osibote O and Balogun V 2022 A methodical review on the applications and potentialities of using nanobiosensors for diseases diagnosis *BioMed Res. Int.* **2022** 1682502

[19] Onyancha R, Ukhurebor K, Aigbe U, Osibote O, Kusuma H, Darmokoesoemo H and Balogun V 2021 A systematic review on the detection and monitoring of toxic gases using carbon nanotube-based biosensors *Sens. Bio-Sens. Res.* **34** 100463

[20] Aigbe U, Ukhurebor K, Onyancha R, Osibote O, Darmokoesoemo H and Kusuma H 2021 Fly ash-based adsorbent for adsorption of heavy metals and dyes from aqueous solution: a review *J. Mater. Res. Technol.* **14** 2751–74

[21] Onyancha R, Aigbe U, Ukhurebor K and Muchiri P 2021 Facile synthesis and applications of carbon nanotubes in heavy-metal remediation and biomedical fields: a comprehensive review *J. Mol. Struct.* **1238** 130462

[22] Kerry R *et al* 2021 A comprehensive review on the applications of nano-biosensor based approaches for non-communicable and communicable disease detection *Biomaterials* **9** 3576–602

[23] Peigneux A, Puentes-Pardo J, Rodríguez-Navarro A, Hincke M and Jimenez-Lopez C 2020 Development and characterization of magnetic eggshell membranes for lead removal from wastewater *Ecotoxicol. Environ. Saf.* **192** 110307

[24] Li Z, Xu S, Xiao G, Qian L and Song Y 2019 Removal of hexavalent chromium from groundwater using sodium alginate dispersed nano zero-valent iron *J. Environ. Manag* **244** 33–9

[25] Ukhurebor K, Aigbe U, Onyancha R, Nwankwo W, Osibote O, Paumo H, Ama O, Adetunji C and Siloko I 2021 Effect of hexavalent chromium on the environment and removal techniques: a review *J. Environ. Manage.* **280** 111809

[26] El-Nemr M, Aigbe U, Ukhurebor K, Onyancha R, El Nemr A, Ragab S, Osibote O and Hassaan M 2022 Adsorption of Cr^{6+} ion using activated pisum sativum peels-triethylene-tetramine *Environ. Sci. Pollut. Res.* **29** 1–25

[27] Terzopoulou A *et al* 2019 Mineralization-inspired synthesis of magnetic zeolitic imidazole framework composites *Angew. Chem. Int. Ed.* **58** 13550–5

[28] Zhao L-X, Xiao H, Li M-H, Xie M, Li N and Zhao R-S 2021 Effectively removing indole-3-butyric acid from aqueous solution with magnetic layered double hydroxide-based adsorbents *J. Hazard. Mater.* **408** 124446

[29] Ranjan B, Pillai S, Permaul K and Singh S 2019 Simultaneous removal of heavy metals and cyanate in a wastewater sample using immobilized cyanate hydratase on magnetic-multi-wall carbon nanotubes *J. Hazard. Mater.* **363** 73080

[30] Huang X, Xu C, Ma J and Chen F 2018 Ionothermal synthesis of Cu-doped Fe_3O_4 magnetic nanoparticles with enhanced peroxidase-like activity for organic wastewater treatment *Adv. Powder Technol.* **29** 796–803

[31] Arabkhani P, Javadian H, Asfaram A and Ateia M 2021 Decorating graphene oxide with zeolitic imidazolate framework (ZIF-8) and pseudo-boehmite offers ultra-high adsorption capacity of diclofenac in hospital effluents *Chemosphere* **271** 129610

[32] Dil E, Ghaedi M, Asfaram A and Mehrabi F 2017 Application of modified magnetic nanomaterial for optimization of ultrasound-enhanced removal of Pb^{2+} ions from aqueous solution under experimental design: investigation of kinetic and isotherm *Ultrason. Sonochem.* **36** 409–19

[33] Rajput S, Singh L, Pittman C and Mohan D 2017 Lead (Pb^{2+}) and copper (Cu^{2+}) remediation from water using superparamagnetic maghemite (γ-Fe_2O_3) nanoparticles synthesized by flame spray pyrolysis (FSP) *J. Colloid Interface Sci.* **492** 176–90

[34] Sahoo S, Padhiari S, Biswal S, Panda B and Hota G 2020 Fe_3O_4 nanoparticles functionalized GO/g-C_3N_4 nanocomposite: an efficient magnetic nanoadsorbent for adsorptive removal of organic pollutants *Mater. Chem. Phys.* **244** 122710

[35] Shukla S, Khan R and Daverey A 2021 Synthesis and characterization of magnetic nanoparticles, and their applications in wastewater treatment: a review *Environ. Technol. Innov* **24** 101924

[36] Das T, Ganguly S, Ghosh S, Remanan S, Ghosh S and Das N 2019 In-situ synthesis of magnetic nanoparticle immobilized heterogeneous catalyst through mussel mimetic approach for the efficient removal of water pollutants *Colloid Interface Sci. Commun.* **33** 100218

[37] Rawat D, Kumari A and Singh R 2021 Synthesis and functionalization of magnetic and semiconducting nanoparticles for catalysis *Functionalized Nanomaterials for Catalytic Application* (New York: Wiley) 261–302

[38] Kazemi M 2020 Based on magnetic nanoparticles: gold reusable nanomagnetic catalysts in organic synthesis *Synth. Commun.* **50** 2079–94

[39] Zhang Q, Yang X and Guan J 2019 Applications of magnetic nanomaterials in heterogeneous catalysis *ACS Appl. Nano Mater.* **2** 4681–97

[40] Leslie-Pelecky D and Rieke R 1996 Magnetic properties of nanostructured materials *Chem. Mater.* **8** 770–1783

[41] Xu Z, Hou Y and Sun S 2007 Magnetic core/shell Fe_3O_4/Au and Fe_3O_4/Au/Ag nanoparticles with tunable plasmonic properties *J. Am. Chem. Soc.* **129** 8698–9

[42] Zeng H, Li J, Wang Z, Liu J and Sun S 2004 Bimagnetic core/shell FePt/Fe_3O_4 nanoparticles *Nano Lett.* **4** 187–90

[43] Wang P *et al* 2016 Silica coated Fe$_3$O$_4$ magnetic nanospheres for high removal of organic pollutants from wastewater *Chem. Eng. J.* **306** 280–8

[44] Shao M, Ning F, Zhao J, Wei M, Evans D and Duan X 2012 Preparation of Fe$_3$O$_4$@SiO$_2$@layered double hydroxide core–shell microspheres for magnetic separation of proteins *J. Am. Chem. Soc.* **134** 1071–7

[45] Pati S, Singh L, Guimarães E, Mantilla J, Coaquira J, Oliveira A, Sharma V and Garg V 2016 Magnetic chitosan-functionalized Fe$_3$O$_4$@Au nanoparticles: synthesis and characterization *J. Alloys Compd.* **684** 68–74

[46] Kong L, Lu X, Bian X, Zhang W and Wang C 2011 Constructing carbon-coated Fe$_3$O$_4$ microspheres as antiacid and magnetic support for palladium nanoparticles for catalytic applications *ACS Appl. Mater. Interfaces* **3** 35–42

[47] Zhu T, Chen J and Lou X 2011 Glucose-assisted one-pot synthesis of FeOOH nanorods and their transformation to Fe$_3$O$_4$@carbon nanorods for application in lithium ion batteries *J. Phys. Chem. C* **115** 9814–20

[48] Deng D, Shen Y, Zhu H and Tu T 2017 A magnetic nanoparticle-supported N-heterocyclic carbene-palladacycle: an efficient and recyclable solid molecular catalyst for Suzuki–Miyaura cross-coupling of 9-chloroacridine *Chem. Commun.* **53** 13063–6

[49] Polshettiwar V, Baruwati B and Varma R 2009 Nanoparticle-supported and magnetically recoverable nickel catalyst: a robust and economic hydrogenation and transfer hydrogenation protocol *Green Chem.* **11** 127–31

[50] Yao Y, Rubino S, Gates B, Scott R and Hu Y 2017 *In situ* x-ray absorption spectroscopic studies of magnetic Fe@FexOy/Pd nanoparticle catalysts for hydrogenation reactions *Catal. Today* **291** 180–6

[51] Dinamarca R, Espinoza-González R, Campos C and Pecchi G 2019 Magnetic Fe$_2$O$_3$–SiO$_2$–MeO$_2$–Pt (Me = Ti, Sn, Ce) as catalysts for the selective hydrogenation of cinnamaldehyde. Effect of the nature of the metal oxide *Materials* **12** 413

[52] Socas-Rodríguez B, Herrera-Herrera A, Asensio-Ramos M and Rodríguez-Delgado M A 2020 Recent applications of magnetic nanoparticles in food analysis *Processes* **8** 1140

[53] Ulusoy H, Gülle S, Yilmaz E and Soylak M 2019 Trace determination of vitamin B12 in food samples by using Fe$_3$O$_4$ magnetic particles including multi-walled carbon nanotubes and nanodiamonds *Anal. Methods* **11** 5108–17

[54] Sereshti H, Khodayari F and Nouri N 2019 Simultaneous determination of aflatoxins in bread by in-syringe dispersive micro-solid phase extraction using magnetic three-dimensional graphene followed by HPLC-FLD *Food Anal. Methods* **12** 2273–81

[55] Manousi N, Giannakoudakis D, Rosenberg E and Zachariadis G A 2019 Extraction of metal ions with metal–organic frameworks *Molecules* **24** 4605

[56] Wang P-L, Xie L-H, Joseph E, Li J-R, Su X-O and Zhou H-C 2019 Metal–organic frameworks for food safety *Chem. Rev.* **119** 10638–90

[57] Liu J, Li G, Wu D, Yu Y, Chen J and Wu Y 2020 Facile preparation of magnetic covalent organic framework–metal organic framework composite materials as effective adsorbents for the extraction and determination of sedatives by high-performance liquid chromatography/tandem mass spectrometry in meat samples *Rapid Commun. Mass Spectrom.* **34** e8742

[58] Shakourian M, Yamini Y and Safari M 2020 Facile magnetization of metal–organic framework TMU-6 for magnetic solid-phase extraction of organophosphorus pesticides in water and rice samples *Talanta* **218** 121139

[59] Durmus Z, Zengin Kurt B, Gazioğlu I, Sevgi E and Kizilarslan Hancer C 2020 Spectrofluorimetric determination of aflatoxin B1 in winter herbal teas via magnetic solid phase extraction method by using metal–organic framework (MOF) hybrid structures anchored with magnetics nanoparticles *Appl. Organomet. Chem.* **34** e5375

[60] Senosy I, Guo H-M, Ouyang M-N, Lu Z-H, Yang Z-H and Li J-H 2020 Magnetic solid-phase extraction based on nano-zeolite imidazolate framework-8-functionalized magnetic graphene oxide for the quantification of residual fungicides in water, honey and fruit juices *Food Chem.* **325** 126944

[61] Kusuma H, Ukhurebor K, Aigbe U, Onyancha R, Onyeachu I and Darmokoesoemo H 2023 Role of magnetic nanomaterials in biosafety and bioregulation facets *Magnetic Nanomaterials. Engineering Materials* (Berlin: Springer Nature) pp 217–34

[62] Adetunji C, Ukhurebor K, Olaniyan O, Olugbenga S, Oloke J and Ubie B 2022 Ethical and social aspects of modern biotechnology *Biosafety and Bioethics in Biotechnology* (New York: CRC Press) pp 51–67

[63] Nwankwo W, Adetunji C, Ukhurebor K and Makinde S 2023 Artificial intelligence-aided bioengineering of eco-friendly microbes for food production policy and security issues in a developing society *Agricultural Biotechnology: Food Security Hot Spots* (New York: Taylor and Francis) pp 301–14

[64] Adetunji C, Nwankwo W, Makinde S, Ukhurebor K, Olaniyan O, Masajuwa F and Ubi B 2022 The role of an intelligent feedback control system in the standardization of bio-fermented food products *Fermentation and Algal Biotechnologies for the Food, Beverage and Other Bioproduct Industries* (New York: CRC Press) pp 143–62

[65] Adetunji C, Olugbemi O, Anani O, Hefft D, Nwankwo W, Olayinka A and Ukhurebor K 2022 Cyberespionage: socioeconomic implications on sustainable food security *AI, Edge and IoT-based Smart Agriculture* (Elsevier: Academic) pp 437–47

[66] Ukhurebor K and Aidonojie P 2021 The Influence of climate change on food innovation technology: review on topical developments and legal framework *Agri. Food Secur.* **10** 1–14

[67] Ukhurebor K, Mishra P, Mishra R and Adetunji C 2020 Nexus between climate change and food innovation technology: recent advances *Innovations in Food Technology* (Berlin: Springer Nature) pp 289–99

[68] Anani A, Adama K, Ukhurebor K, Habib A, Abanihi V and Pal K 2023 Application of nanofibrous protein for the purification of contaminated water as a next generational sorption technology: a review *Nanotechnology* **34** 1–18

[69] Ukhurebor K, Hossain I, Pal K, Jokthan G, Osang F, Ebrima F and Katal D 2023 Applications and contemporary issues with adsorption for water monitoring and remediation: a facile review *Top. Catal.* **67** 140–55

[70] Aidonojie P, Ukhurebor K, Oaihimire I, Ngonso B, Egielewa P, Akinsehinde B, Heri S and Darmokoesoemo H 2023 Bioenergy revamping and complimenting the global environmental legal framework on the reduction of waste materials: a facile review *Heliyon* **9** e12860

[71] Tsuzuki T 2013 *Nanotechnology Commercialisation* (New York: Taylor and Francis)

[72] ISO 2003 *ISO 14040-1997, Environmental Management—Life Cycle Assessment-Principles and framework* (Geneva: ISO)

[73] Pallas G, Vijver M G, Peijnenburg W J G M and Guinée J 2020 Life cycle assessment of emerging technologies at the lab scale: the case of nanowire-based solar cells *J. Ind. Ecol.* **24** 193–204

[74] Mullen E and Morris M 2021 Green nanofabrication opportunities in the semiconductor industry: a life cycle perspective *Nanomaterials* **11** 1085

[75] Feijoo S, González-García S, Moldes-Diz Y, Vazquez-Vazquez C, Feijoo G and Moreira M 2017 Comparative life cycle assessment of different synthesis routes of magnetic nanoparticles *J. Clean. Prod.* **143** 528–38

[76] Baresel C, Schaller V, Jonasson C, Johansson C, Bordes R, Chauhan V, Sugunan A, Sommertune J and Welling S 2019 Functionalized magnetic particles for water treatment *Heliyon* **5** e02325

[77] Marimón-Bolívar W and González E 2018 Green synthesis with enhanced magnetization and life cycle assessment of Fe3O4 nanoparticles *Environ. Nanotechnol. Monit. Manag* **9** 58–66

[78] Lauria A and Lizundia E 2020 Luminescent carbon dots obtained from polymeric waste *J. Clean. Prod.* **262** 121288

[79] Celik G *et al* 2019 Upcycling single-use polyethylene into high-quality liquid products *ACS Cent. Sci.* **5** 1795–803

[80] Joshi N, Filip J, Coker V, Sadhukhan J, Safarik I and Bagshaw H L J 2018 Microbial reduction of natural Fe(III) minerals; toward the sustainable production of functional magnetic nanoparticles *Front. Environ. Sci.* **6** 127

[81] Nana L, Ruiyi L, Qinsheng W, Yongqiang Y, Xiulan S, Guangli W and Zaijun L 2021 Colorimetric detection of chlorpyrifos in peach based on cobalt-graphene nanohybrid with excellent oxidase-like activity and reusability *J. Hazard. Mater.* **415** 125752

[82] Warheit D, Webb T, Sayes C, Colvin V and Reed K 2006 Pulmonary instillation studies with nanoscale TiO_2 rods and dots in rats: toxicity is not dependent upon particle size and surface area *Toxicol. Sci.* **91** 227–36

[83] Roy S, Das K and Dhar S 2021 Conventional to green synthesis of magnetic iron oxide nanoparticles; its application as catalyst, photocatalyst and toxicity: a short review *Inorg. Chem. Commun.* **134** 109050

[84] Malhotra N, Lee J-S, Liman R, Ruallo J, Villaflores O, Ger T-R and Hsiao C- D 2020 Potential toxicity of iron oxide magnetic nanoparticles: a review *Molecules* **25** 3159

[85] Soenen S, Rivera-Gil P, Montenegro J-M, Parak W, De Smedt S and Braeckmans K 2011 Cellular toxicity of inorganic nanoparticles: common aspects and guidelines for improved nanotoxicity evaluation *Nano Today* **6** 446–65

[86] Mahmoudi M, Simchi A, Milani A and Stroeve P 2009 Cell toxicity of superparamagnetic iron oxide nanoparticles *J. Colloid Interface Sci.* **336** 510–8

[87] Hafiz A, Hassanein M, Talhami M, AL-Ejji M, Hassan A and Hawari A 2022 Magnetic nanoparticles draw solution for forward osmosis: current status and future challenges in wastewater treatment *J. Environ. Chem. Eng.* **10** 108955

[88] Ran Q, Xiang Y, Liu Y, Xiang L, Li F, Deng X, Xiao Y Y, Chen L, Chen L and Li Z 2015 Eryptosis indices as a novel predictive parameter for biocompatibility of Fe_3O_4 magnetic nanoparticles on erythrocytes *Sci. Rep.* **5** 16209

[89] Zhu X, Tian S and Cai Z 2012 Toxicity assessment of iron oxide nanoparticles in zebrafish (*Danio rerio*) early life stages *PLoS One* **7** e46286

[90] Shakil M, Hasan M, Uddin M, Islam A, Nahar A, Das H, Khan M, Dey B, Rokeya B and Hoque S 2020 *In vivo* toxicity studies of chitosan-coated cobalt ferrite nanocomplex for its application as MRI contrast dye *ACS Appl. Bio Mater.* **3** 7952–64

[91] Aibani N, Rai R, Patel P, Cuddihy G and Wasan E 2021 Chitosan nanoparticles at the biological interface: implications for drug delivery *Pharmaceutics* **13** 1686

[92] Cole A, David A, Wang J, Galbán C and Yang V 2011 Magnetic brain tumor targeting and biodistribution of long-circulating PEG-modified, cross-linked starch-coated iron oxide nanoparticles *Biomater.* **32** 6291–301

[93] Erdem M, Yalcin S and Gunduz U 2017 Folic acid-conjugated polyethylene glycol-coated magnetic nanoparticles for doxorubicin delivery in cancer chemotherapy: preparation, characterization, and cytotoxicity on HeLa cell line *Hum. Exp. Toxicol.* **36** 833–45

[94] NRC 1983 *National Research Council: Risk Assessment in the Federal Government: Managing the Process* (Washington, DC: National Academies Press)

[95] Torres J and Bobst S 2015 Introduction *Toxicological Risk Assessment for Beginners* (Cham: Springer International Publishing)

[96] Bette Meek M *et al* 2013 A framework for fit-for-purpose dose response assessment *Regul. Toxicol. Pharmacol.* **66** 234–40

[97] Econmarketresearch 2023 *Magnetic Nanoparticles Market Size, Growth & Forecast 2031* (Econmarketresearch)

[98] Precisionbusinessinsights 2023 *Magnetic Nanoparticles Market Size, Share, Growth Analysis* (Precisionbusinessinsights)

[99] Onyancha R, Aigbe U, Ukhurebor K, Kusuma H, Darmokoesoemo H, Osibote O and Pal K 2022 Influence of magnetism-mediated potentialities of recyclable adsorbents of heavy metal ions from aqueous solutions—an organized review *Res. Chem.* **4** 100452

[100] Aigbe U, Onyancha R, Ukhurebor K and Obodo K 2020 Removal of fluoride ions using polypyrrole magnetic nanocomposite influenced by rotating magnetic field *RSC Adv.* **10** 595–609

[101] Zhu Y, Stubbs L, Ho F, Liu R, Ship C, Maguire J and Hosmane N 2010 Magnetic nanocomposites: a new perspective in catalysis *ChemCatChem.* **2** 365–74

[102] Polshettiwar V, Luque R, Fihri A, Zhu H H, Bouhrara M and Basset J-M 2011 Magnetically recoverable nanocatalysts *Chem. Rev.* **111** 3036–75

[103] Oliveira R, Zanchet D, Kiyohara P and Rossi L 2011 On the stabilization of gold nanoparticles over silica-based magnetic supports modified with organosilanes *Chem. Eur. J.* **17** 4626–31

[104] Ge W, Encinas A, Araujo E and Song S 2017 Magnetic matrices used in high gradient magnetic separation (HGMS): a review *Res. Phys.* **7** 4278–86

[105] Stephens J, Beveridge J and Williams M 2012 Analytical methods for separating and isolating magnetic nanoparticles *Phys. Chem. Chem. Phys.* **14** 3280–9

[106] Reddy P, Chang K-C, Liu Z-J, Chen C-T and Ho Y-P 2014 Functionalized magnetic iron oxide (Fe_3O_4) nanoparticles for capturing gram-positive and gram-negative bacteria *J. Biomed. Nanotechnol.* **10** 1429–39

[107] Ditsch A, Lindenmann S, Laibinis P, Wang D and Hatton T 2005 High-gradient magnetic separation of magnetic nanoclusters *Ind. Eng. Chem. Res.* **44** 6824–36

[108] Hasani M and Jafari A 2022 Electromagnetic field's effect on enhanced oil recovery using magnetic nanoparticles: microfluidic experimental approach *Fuel* **307** 121718

[109] Ri H-C, Jon C-S, Lu L, Piao X and Li D 2022 A dynamic electromagnetic field assisted boronic acid-modified magnetic adsorbent on-line extraction of cis-diol-containing flavonoids from onion sample *J. Food Compos. Anal.* **119** 105279

[110] Dadfar S, Kögerler P, Hermanns-Sachweh B, Schulz V, Kiessling F, Lammers T *et al* 2020 Size-isolation of superparamagnetic iron oxide nanoparticles improves MRI, MPI and hyperthermia performance *J. Nanobiotechnol.* **18** 22

[111] Mrówczyński R, Nan A and Liebscher J 2014 Magnetic nanoparticle-supported organo-catalysts—an efficient way of recycling and reuse *RSC Adv.* **4** 5927–52

[112] Rezaei H, Shahbazi K and Behbahani M 2021 Application of amine modified magnetic nanoparticles as an efficient and reusable nanofluid for removal of Ba^{2+} in high saline waters *Silicon* **13** 4443–51

[113] Hedayatnasab Z, Abnisa F and Daud W 2017 Review on magnetic nanoparticles for magnetic nanofluid hyperthermia application *Mater. Des.* **123** 174–96

[114] Karimi M *et al* 2016 Smart micro/nanoparticle in stimulus-responsive drug/gene delivery systems *Chem. Soc. Rev.* **45** 1457–501

[115] Ulbrich K, Holá K, Šubr V, Bakandritsos A, Tuček J and Zbořil R 2016 Targeted drug delivery with polymers and magnetic nanoparticles: covalent and noncovalent approaches, release control, and clinical studies *Chem. Rev.* **116** 5338–431

[116] Long N, Yang Y, Teranishi T, Thi C, Cao Y and Nogami M 2015 Related magnetic properties of $CoFe_2O_4$ cobalt ferrite particles synthesised by the polyol method with $NaBH_4$ and heat treatment: new micro and nanoscale structures *RSC Adv.* **5** 56560–9

www.ingramcontent.com/pod-product-compliance
Lightning Source LLC
Chambersburg PA
CBHW080548220326

41599CB00032B/6401